THE
FIELDCRAFT
OF
LEADERSHIP

BY KOREY STALEY

AUTHOR'S NOTE

In the interest of maintaining the privacy and safety of individuals, as well as the security of operational details, certain names and aspects of military missions described within this book have been altered. These changes have been made carefully to preserve the integrity of the recounted events and the experiences of those who lived them. While the essence and the historical accuracy remain untouched, it is my aim to share these compelling stories with respect and diligence, honoring those who have served while safeguarding certain information.

In this book, there are several examples of poor leadership. If you come across a story and believe you are being cited as an example and find the portrayal of your character less than flattering, get over it. We all have made mistakes, me included. Hopefully, you have learned from yours, just as I continue to learn from mine.

Throughout my twenty-two-year military career, I worked closely with hundreds, if not thousands, of soldiers, and the vast majority were male. This was simply the nature of being in combat units, where we spent nearly all our time deployed in remote operating bases. If I tend to speak primarily in the context of men, it's because that was my predominant experience. Of course, the principles I discuss apply universally. This book isn't meant to be exclusionary; it just happens to be my own experience.

Foreword

CPT (Ret) Florent Groberg

CONGRESSIONAL MEDAL OF HONOR RECIPIENT

I consider it an honor and privilege to be writing the foreword for a book that encapsulates the wisdom and leadership lessons garnered from the trenches of the battlefield. This book is not just about leadership. It is about humanity, courage, grit, and the indomitable spirit of individuals who have faced the harshest realities and yet held onto their sense of duty and purpose.

I have had the unique fortune of knowing and learning from Korey Staley, the author of this invaluable guide. I first met him on the battlefronts of Afghanistan, where his leadership was a beacon for all the soldiers under his command. Despite being under constant threat, he was unyielding in his resolve to guide his men through the battlefield, making strategic decisions that demanded not just an understanding of warfare but also the intricacies of human nature.

In the battlefield, where life and death decisions are made within the blink of an eye, Staley's leadership style shone through. He was not just focused on achieving the mission objectives, but he was also deeply invested in the welfare of his men. He embodied the principle of "taking care of your boys," a lesson he held dear throughout his service. For him, his men were not just pawns in a larger game of war; they were individuals with fears, hopes, and dreams—a reality that he never lost sight of, even under the most challenging circumstances.

Staley is a master craftsman of leadership, having honed his skills through years of hands-on experience in one of the most challenging professions in the world. His lessons are not theoretical constructs taught in a classroom, they are hard-earned insights from the school of life. His techniques, strategies, and wisdom were shaped through his trials and tribulations, making them invaluable for anyone aspiring to be a leader, in or out of the military.

As I read through this book, I found myself reminiscing about my early days in the military, especially the time I arrived at Forward Operating Base Honaker Miracle, Afghanistan. As a fresh Lieutenant, eager to prove myself but also aware of my green status, I sought Staley for guidance. He didn't just offer me advice, he provided me with a roadmap to understand, learn, and ultimately lead in an environment as unforgiving as the battlefield. As a brand-new leader, he could have easily made life difficult for me, but he decided to teach me instead.

His words still echo in my mind, "Shut up for the next seven days." I took his advice to heart, becoming a keen observer of the actions and reactions of those around me. I immersed myself in the dynamics of the battlefield, observing the seamless coordination, quick decisions, and instinctual actions that could mean the difference between life and death. His method of silent learning was unorthodox, but it provided invaluable insights into leadership that a conventional lecture could never have provided.

This book, much like Staley's leadership, is not about grand theories. It's about pragmatic, ground-level leadership principles. It's about what truly matters in leadership: taking care of your team, understanding the ground realities, building relationships, earning trust, and, most importantly, leading by example.

The Fieldcraft of Leadership is a testament to the journey of a man who led not because he had to but because he wanted to. His purpose was not just to guide his men through a mission but to ensure their safe return. His dedication to his team, his commitment to the mission, and his courage to make tough decisions are the very embodiments of true leadership.

This book is not just a guide, it is a tribute to the many unsung heroes who have led with dignity and courage under extraordinary circumstances. Korey Staley was a warrior. But more than that, he was a genuine leader. I can't help but marvel at the courage, dedication, and wisdom he displayed throughout his career. His approach to leadership wasn't about being in control, it was about empowering others, fostering trust, and building relationships. He understood the importance of preserving and passing down the fieldcraft of leadership, the unwritten techniques and wisdom that come from real-world experience.

Within these pages, you will find a treasure trove of wisdom, insights, and experiences gleaned from a lifetime of service. You'll learn the raw truth about leading in the real world, about making decisions when the stakes are high and lives are on the line.

In an era where leadership is often misunderstood and misrepresented, Korey Staley offers us a refreshing and powerful perspective. He reminds us that leadership is not about titles or positions, but about the difference we make in the lives of those we lead.

As you read this book, I hope you will absorb the lessons Korey imparts, apply them in your leadership journey, and continue the tradition of passing down this invaluable fieldcraft. These lessons about leadership are timeless and universally applicable, whether you are in the military, in business, or in any field where leading people matters.

I invite you to embark on this enlightening journey. May the lessons you learn inspire and guide you as they have inspired and guided me, and may you pass on the fieldcraft of leadership to future generations.

The world needs more leaders like Korey. Leaders who are willing to put their ego aside for the benefit of the team. Leaders who understand that their role is not to dominate, but to serve. And leaders who realize that the ultimate goal is not just to lead, but to create more leaders.

Contents

Introduction

THE YEAR 2004 CHANGED MY LIFE FOREVER. The decade prior was a relentless marathon of military training. I lived and breathed the rigorous preparation in the forests, mountains, and deserts across the world. Imagine attending daily basketball practice, constantly honing your skills, yet never experiencing the adrenaline of a real game. That was my reality, until 2004. That year I embarked on my first deployment to Iraq. With a blend of anticipation and naive eagerness, I wondered if my exhaustive training would translate into real-world ability. Northern Iraq, a stone's throw from the Syrian border, was a testing field where theory would meet the unvarnished truth of war.

In training, unlike in real combat, bad decisions usually don't carry deadly consequences. Though we strive for realism, most risks in training can be mitigated through meticulous planning and stringent control measures. Yet despite our best efforts, the raw intensity of war cannot be replicated. Combat is an unyielding reality—dangerous, deadly. The enemy gets an equal chance at survival. The enemy occasionally catches us off guard, especially when we're new to a theater of operations. With time and experience, the enemy's playbook starts to read less like a mystery novel and more like a worn-out magazine in a dentist's waiting room—predictable, yet occasionally surprising.

Sun Tzu, in his timeless book *Art of War*, asserts, "If you know the enemy and know yourself, you need not fear the result of a hundred battles." This mantra resonated with me, especially as a newcomer in the unpredictable regions of Iraq. Our knowledge came from aged reports and seasoned stories from preceding units, each one serving to acquaint us with the unfamiliar terrain.

During my first combat deployment as a staff sergeant leading an eleven-man infantry squad, we faced our inaugural major operation in the fall of 2004. Our mission: to raid the stronghold of a notorious IED maker and leader of a local Al Qaeda cell. Despite my brushes with smaller skirmishes, this operation was our squad's dive into deeper waters. In the lead-up, we engaged in detailed planning, aligning with locals and advisors seasoned by nearly a year in the field. Their insights into the enemy's go-to tactics were invaluable. While rehearsing, my mind was preoccupied with thoughts about the unpredictability of the mission outcome. I really felt the weight of leadership when men's lives are at stake. How do I navigate that ultimate responsibility?

One advisor stood out: Sergeant First Class Ashford. His combat experience was as evident as the sun-faded hues of his uniform and the lean look of a man familiar with the rigors of continuous combat. I approached him for guidance. My question was both about the specifics of our mission and the broader challenges of the coming year. His simple advice struck a chord and served as a reminder that amid the complexities of war, the most important thing was to: **"Just focus on taking care of your boys the best you can. Everything else will fall into place."**

The missions, firefights, and shitty days are inevitable in war; perfection is a myth. There is nothing you can do about it. No amount of planning can guarantee one hundred percent success. There are going to be many difficulties you must deal with. Stress comes from the battlefield, but also from home. Soldiers will miss birthdays and holidays. There will be stress about finances or infidelity. On top of that, there's a chance they will have to deal with their closest friends being killed in combat and will still need to function. Above all else, you must be there for each other.

Sergeant First Class Ashford had become hardened through his years in combat. He had experienced the harshest realities of war and figured out ways to adapt and survive. By continually learning and refining techniques, he developed ways that would give his team the best chances of successfully accomplishing the mission and surviving. Despite the simplicity, I was lucky to receive his clarity that day. His advice encapsulates the essence of leadership in such extreme conditions. I used it through many more deployments.

The burden and privilege of leading men in combat was now mine.

This book summarizes the finest leadership qualities, techniques, and traits I learned throughout my twenty-two-year military career. My education in leadership spanned many formal military schools, including Airborne School, Ranger School, Primary Leadership Development Course, and numerous Noncommissioned Officer Courses. Yet the most profound lessons can't be found in a syllabus. They emerged from the raw experiences of observing my leaders and peers in action—particularly in combat's unforgiving hold. As you delve into this book, you will discover that some leadership techniques and traits seem deceptively simple, even rooted in common sense. Yet it's often these overlooked simplicities that harbor the greatest power.

I served in the US Army as an Infantryman from 1993-2016. During that time, I performed in a variety of positions from Long Range Surveillance, Reconnaissance Scout, Sniper, Sniper Section Leader, Team Leader, Squad Leader, Drill Sergeant, Platoon Sergeant, and First Sergeant. I deployed several times culminating a total of thirty-eight months spent in combat. It was a few years after my retirement from the service when I was able to reflect more clearly on what I had learned and how to implement those skills, that I decided to consolidate a list of the most significant lessons and why each is so important.

Military leadership education in schools, academies, or courses often skirted around the art of influencing your team. Instead, their curriculum revolved around the mechanics of leadership—the logistics of equipping your team, ensuring their readiness. Depending on your field, you might dive into the tactical intricacies of ambushes, raids, or airfield seizures. These elements are vital in the physical readiness and training of your team. Yet, the craft of building relationships is absent — the foundation that binds a team. They fall short of teaching the essence of earning trust and respect. Without these, your team remains a collection of individuals. Uniting the team under a banner of mutual trust and respect is the mission of true leadership.

The Mastery of Craft

Countless trades have been fine-tuned, tweaked, and rehearsed to the point of mastery. People become proficient at so many things through trial and error, that they eventually become subject matter experts of their craft. Mastery consists of techniques and knowledge that were not necessarily taught in school. People acquire it over time, through sheer experience. It demands more than textbook knowledge; it requires skills passed down from generation to generation, at each stage refining the craft until the technique is mastered.

Consider the blacksmith. Think about all the techniques and skills required to develop mastery of this craft, which were honed over centuries. Knowing the delicacies of temperatures, the makeup of different metals, when to use each hammer, and the precise moment to strike is an art. To become the master blacksmith, you have to start as an apprentice. Today we go to school to learn the basics of any trade. But for many roles, like becoming a surgeon, you do not become an expert by just studying in school, you need to practice. You need to work with mentors who are more experienced. You need to learn the subtle techniques of how to be an expert from the men and women who came before you.

In the military, the term "fieldcraft" has come to describe things we do to make harsh and hostile environments more tolerable and survivable. Fieldcraft includes the little things shared by leaders with subordinates. These tips and tricks can mean a world of difference in survival on the battlefield or in the woods.

A straightforward military example of fieldcraft is the art of constructing a *hooch*. A hooch is essentially a modest A-frame tarp shelter nestled in the woods, designed to protect from downpours or to provide crucial overhead camouflage. Through a blend of personal trials and the wisdom of our leaders, we've learned a vital technique: if the terrain permits, always position your hooch on a gentle slope. Then craft a suitably sized trench around it. This isn't just busy work. In a torrential downpour, this trench becomes your savior, channeling the flood around the shelter. Without this, the shelter becomes engulfed with water brutally destroying all the time and effort spent constructing it. So, we learn to dig this modest trench, subtly yet effectively rerouting the water. It's a small but crucial piece of knowledge that underscores the importance of detail in survival.

Through my twenty-two years in the military, there have been countless bits of fieldcraft imparted to me. Some of the seemingly smallest tips have turned out to be the best ones. Not all of it was verbal, much of it was through example. I was extremely lucky in my military career to have amazing leaders to learn from. Mentors who took me under their wing and did their best to teach me what it means to be a great leader. I could not always understand and implement that training right away. Despite knowing the techniques, it would sometimes be years before I really understood their significance and how to implement them. One of the reasons it took so long to write this book is the time it has taken me to grasp and comprehend all the leadership roles throughout my life. It took me years in and beyond military service until I was able to fully reflect on what kind of a leader I was striving to be. What I did well and what I could have done differently. Suffice it to say I wasn't perfect by any stretch, and I continue to grow my skills all the time.

Unfortunately, much fieldcraft is lost through time. Sadly, there are skills we will never be able to do as well as our ancestors, at least not without going through a process of trial and error all over again. There is a lot we could learn from the trials of those who came before us.

The Mission of the Book

My mission with this book is to capture the leadership skills, techniques, and knowledge taught to me in the military, as well as the lessons I had to figure out myself. I hope to bridge some of the gaps in knowledge so the learning curve will not be as steep for others, and to add to the proficiency of experienced leaders. It's my hope these skills are top of mind and applicable to almost all situations that leaders may face. Some content in this book is simple, yet I guarantee there will be times in your life or career (be it military or civilian) when you'll need to be reminded to apply it. I hope that by the time you finish reading this book, you become a stronger leader, with a stronger team, in a stronger organization.

I stand with this knowledge on the shoulders of giants.

The leadership techniques in this book must be understood and put into everyday practice, which is why I consolidated them in the last chapter of this book as cliff notes. Review them daily. Ask yourself; *Am I following these principles? How can I implement them?*

Let's get started.

1

Trust and Respect

*"It takes twenty years to build a reputation
and five minutes to ruin it."*
– WARREN BUFFETT

TRUST AND RESPECT ARE THE BEDROCK of leadership. While each chapter in this book stands alone in its value, collectively they are a powerhouse for cultivating the relationships that are crucial for leadership. As you explore deeper through these pages and embrace the guidelines, you will build the necessary skills for leading a team in military fatigues or a civilian suit. Begin by asking yourself these questions: *Does my team follow me because of my leadership abilities, or am I leading primarily from a position of power granted by my title?*

If you rely on your rank to lead, the result will be a dysfunctional and incohesive team. Too often rank is used as a crutch, and you should lead as if you have none. It gets used as a shield, deflecting feedback that you need to hear but are protecting yourself from. You want your team to follow because they trust and respect you, not because you demand it.

I once had a platoon leader, Lieutenant Clark, a man who believed his rank was the only authority he needed. I was a young sergeant, but seasoned enough to sense the shit show that would ensue amongst the team with the Lieutenant's approach.

His leadership style was a stark contrast to what we were accustomed to. In our world, trust was earned, not given with the stripes or bars on our uniforms. But for Lieutenant Clark, his bars were his only credential or at least that was all he was willing to demonstrate.

The first sign of trouble came during a routine training mission. We were navigating through dense woodland. Clark halted the platoon and made a sudden call to change our planned route to one we had not studied. None of us knew why; he didn't bother explaining. I could see the confusion in my squad's eyes, the unspoken questions. Why are we diverting?

His decisions were like that, abrupt and shrouded in mystery. And whenever anyone tried to offer insight, he'd dismiss it with a wave of his hand. "This is my patrol," he'd say. As if those words were enough to shift any doubt.

But doubts grew like weeds. We were trained to adapt and to overcome. Blindly following orders can be a death sentence. His refusal to take advice from more experienced soldiers, especially in complex tactical situations, didn't sit well with us. It wasn't simply a matter of making decisions; it was ensuring they were the right ones, made for the right reasons.

That evening, after what ended up being a particularly grueling day, I approached him. "Lieutenant, we're confused about your last-second decision to change the route. Maybe a little insight from you could help."

He looked at me, indifferently. "Listen, I do not...have to explain...every decision...that I make."

An officer shouldn't have to stop and verbally justify every decision, but there is a balance. His authoritarian approach was chipping away at the foundation of our unit.

In the field, trust isn't a nice-to-have; it's a lifeline. It's knowing the person leading you understands the gravity of every decision. But with Clark, each order, no matter how minor, only deepened our skepticism. It wasn't just a matter of disagreeing with his choices; the issue was his refusal to explain them, instead using his rank as a shield.

Lieutenant Clark never changed his ways. Eventually, his style led to a transfer, which meant he was fired. We got a new platoon leader, one who knew that rank was just a part of leadership, not the whole of it.

Looking back, I realize how crucial it is for leaders to build trust, blend experience with rank, and know that respect is earned by more than just the insignia on your uniform. It's a lesson I carried with me, long after my time under Lieutenant Clark had ended.

During combat, it's even more evident who people listen to and whose lead they will follow. The person who brings the greatest skill, knowledge, and common sense is the person who will influence the team the most.

A CEO, much like a maestro in an orchestra, doesn't need to play every instrument. **Their skill lies in harmonizing the collective expertise of their team.** They are not concerned with playing every note themselves but understanding who in their team can play it best. By respecting everyone's mastery, the CEO creates a symphony of skills, leading to an outstanding performance.

A Meeting with No Rank

Imagine yourself in a meeting with all your subordinates, peers, and even your boss; however, in this meeting, there is no emphasis on rank, title, or position. Each individual is present without any hierarchical label. Everyone is free to voice their opinions at any time. Even the most junior members are encouraged to express their discontent with your ideas, and your peers can call you out for any laziness, bias, incompetence, or disagreeable behavior. Suppose this group is convened to brainstorm and devise a solution to a problem. They must determine how to best tackle it and assign tasks accordingly. If you were to speak up with a potential solution, would people listen to you, and choose your plan? Or would your opinion be brushed aside because you don't have a visible rank on your shirt or a fancy title on your business card? Is there a possibility that they only heed your words because of your position in the hierarchy?

If you were brand new to a team, would you be able to convince your new teammates you are competent and able to lead them toward the best way forward?

When you step into a leadership role, your team may not express their true feelings directly to you. It's in those moments when you're not present, in those candid "no rank" meetings, where genuine opinions surface. Here's the key to deciphering their trust: observe their level of commitment to your ideas. Their effort is a mirror of their respect for you. If they believe your idea lacks merit and they don't hold you in high regard, their effort will reflect it. It's a non-verbal way of saying, *You're an idiot.*

In an ideal world, leaders earn titles through their experience, but this isn't always how it works. Sometimes, you must choose the best person available for a role, even if they lack experience. People respect competence and character. Whether it's you or someone else who's stepping up to lead, expertise will come with time.

Build a Team that Is Willing to Die for You

I stood in front of fifty new lieutenants and platoon sergeants in a large room. The walls were adorned with the history of battles waged by the soldiers of the 17th Infantry Battalion. I was tasked with equipping them for the challenges of leading men into combat. A candid conversation where they could ask questions from the perspective of someone who has led and has been led in combat. These leaders were on the verge of embarking on the most demanding journey of their lives. The air was heavy with anticipation. As a combat-seasoned first sergeant, I was there to offer them the front-line experience and perspective essential for their transition. They were moving from the theoretical learning of an auditorium classroom to the real-world classroom of fighting, chaos, emotions, and the stark reality of death. I started with a clear statement and then posed the most critical question a leader in our situation must consider.

"You must build a team of people willing to risk their lives for you. How will you inspire a team so they're willing to risk everything for each other?"

This isn't just a question; it's a call to awaken the profound sense of purpose that true leadership demands. For these military leaders, this is their reality—creating a bond so strong that team members are willing to put their lives on the line for each other. As the conversation with these young leaders wrapped up, the conclusion was clear: for people to be willing to risk everything, there must be a bond as strong as a family, which is built on the foundation of trust and respect.

Every young soldier joins the military for many different reasons. They join for college money, to see the world, or some don't know what else to do. They join to escape a bad home situation or maybe some just romanticized idea of being in the military and going to war. Honestly, I believe some of them join because they just like playing *Call of Duty* and thought it would be the same. How do you unify individuals from diverse backgrounds into a cohesive team that is ready to risk their lives? In the thick of battle, as bullets whizz by, thoughts of college funds evaporate from the young soldier's mind.

The driving force behind their willingness to risk their lives and put themselves in harm's way is the relationship you have cultivated with them and the team. To instill a sense of camaraderie that enables a soldier to risk everything for another, you must work hard to build and foster that bond. This is no easy task. If you allow toxic people or situations to go unchecked in your organization or team, achieving this vital mindset is impossible. When the chips are down, your team members will not trust you, and they will be unwilling to put their lives at risk for you. Therefore, you must ask yourself: *Are you leading with trust or rank?* Ask this question daily. You cannot rely solely on your rank.

Now, let's translate this to civilian organizations. It's not about the literal risking of lives, but about fostering a culture where each team member is so committed, so driven by a shared mission, that they give their absolute best. How transformative could your organization be with a team that's deeply dedicated to its collective success? Think about the immense power of unwavering commitment. That's the kind of environment you want to create, no matter what your field is.

Rules for Maintaining the Trust of Your Team:

Once trust is lost, it is nearly impossible to earn back. Here are some crucial leadership qualities needed to build and maintain trust.

- **Never lie to your team.** People aren't stupid, and they will see right through it. Assuming a person isn't smart enough to know when you're not telling them the truth is insulting. Be straightforward and transparent. Never bullshit them. They are adults and don't need things to be sugar-coated.

- **Always follow through** with your word. It's a very simple concept, but can be difficult in practice. Never promise anything you can't 100% deliver.

- **Be genuine**. You're in the wrong line of work if you don't care about your troops. If you appear fake in your sincerity, they recognize it.

- **Don't be a pushover**. A leader who asserts the necessary firmness is respected more than one who doesn't. Your team expects you to hold them and others accountable for their actions. Allowing people to walk all over you doesn't gain their favor.

- **Don't overreact or be dramatic.** Your team will see you as being immature and not worthy of leading.

- **No name-calling.** Never speak to anyone in a demeaning way. You may have to be harsh sometimes but never do it in a degrading way. Your job is to lift your team up, not put them down. Don't say things you can't take back. Treat these individuals as if they're family.

- **Get your hands dirty.** It's important to be visibly engaged in day-to-day work when it counts. Don't assume you already have respect just because you had it yesterday. Respect is earned daily, and you should ask yourself: *What have I done today to deserve this position?*

- **Take accountability.** We all make mistakes, but not owning your mistakes destroys your credibility. Take responsibility for your actions as a leader, good or bad. A humble man can be more powerful than a prideful man.

Course of Action Matrix (COA-M)

A COA-M is a method to determine the best way to accomplish a mission. It also is great for building trust and buy-in from the team. As a private, I learned about this concept when I was part of a reconnaissance detachment for the 101st Airborne Division. Our primary job was to conduct surveillance, pinpoint targets behind enemy lines, and relay our findings back to the division intelligence teams. Whenever we were given a target, we started to plan a recon mission.

Each team member independently developed a plan, detailing every aspect of execution. It was a strategic brainstorming session—deciding on insertion methods, landing zones, and plotting a stealthy route to the Objective Rally Point and Hide Site. Then we'd meticulously select the observation spot, striving for the perfect balance of visibility of the objective and concealment from the enemy.

Once each person finalized their plan, they would brief the group. Explaining their strategic thought process for each part of their plan.

Each person's plan was dissected with equal intensity, focusing solely on the merit of the ideas. In these planning sessions, rank took a backseat. From the greenest to the most seasoned sergeant, every voice mattered.

The matrix served as a detailed score breakdown for each plan, with every aspect rated on a scale from one to five. For instance, the group might appreciate the landing site and assign it a score of four. However, if the route was deemed too long or lacked sufficient cover, it would receive a score of three. This evaluation process was applied consistently across all plans until the one with the highest score was identified. Typically, this involved selecting and integrating the most effective elements from various plans to achieve the best overall score.

Generally, the soldiers with the highest rank possessed the greatest depth of knowledge and experience. However, there were times when the most junior soldier provided a suggestion that was the most tactically sound, bringing a fresh perspective to the situation. It was a melding of minds, and the critique was based on the merit of the plan alone.

You can use the COA-M across various planning or brainstorming activities in any field. It can be applied to individual plans or used by groups working collaboratively to formulate a plan. The primary objective is to identify the best possible plan, rather than discrediting any particular plan. This isn't anyone's time to show off or to put others down, so everyone needs to put their ego away.

It's worth noting that many leaders are reluctant to seek input from junior members, and even less inclined to have their own plans critiqued by junior members. If your experience doesn't back up your stripes, your team will see right through it. They know who the real veterans are. Ego has no place here; it's about harnessing everyone's talents for collective success. Recognize and give credit where it is due.

Leadership Challenge

As a leader, you will undoubtedly face moments when your team members disagree with your decisions or actions. They might perceive your approach as overly strict or feel that your instructions clash with their personal interests. Balancing individual preferences with the collective good can be a tough task. It's in these moments that you'll have to make hard choices that might not be well-received. And that's life. While it might be difficult for some to accept, it's entirely feasible to maintain respect while holding differing opinions. Regardless of whether your team agrees, if they respect you, they will implement your plan with the full intent to succeed. Especially if they have buy-in.

Remember, the path to expertise is paved with action. With each step you take and each decision you make, you'll gain invaluable experience, and with it, your team's confidence in you. True leadership isn't just about accumulating knowledge, it's about maintaining humility and keeping your team's welfare at the forefront. When you genuinely prioritize their interests, trust and respect will follow. Let this be your guiding principle and the compass that steers your leadership.

KEY POINTS

- Put your rank aside and lead with respect. Treat everyone the same.

- Own your words—don't lie, don't sugarcoat, and always follow through.

- Get in the trenches and work hard—your actions show what real competence and character look like.

- Bring your team into the fight—let them own the plan and use their strengths to drive success.

- Build real relationships. Treat your team like family—lead with honesty, toughness, and connection.

- The respect and effort your team gives you is a mirror of your leadership—stay composed and balanced, even under pressure.

2
Know Your People

"Everyone you meet knows something you don't."
– BILL NYE

IN THE RUGGED TERRAIN OF AFGHANISTAN, the US military strategy hinged on a concept that was as straightforward as it was complex: win the hearts and minds of the local population. This phrase wasn't just a slogan; it underscored the crucial need to gain both emotional support and trust from the Afghans. We understood that military success was impossible without the locals embracing our efforts—they were the key to rejecting the Taliban.

Prior to deployment, our preparation extended beyond rehearsing raids, ambushes, and doing endless pushups. We also engaged in cultural awareness classes—an experience both enlightening and amusing. Imagine a group of hardened soldiers ready for combat, receiving classes on navigating the subtleties of Afghan customs. Simple courtesies such as entering a room ahead of an elder is a sign of disrespect. Should we inadvertently piss off the local people, there's a risk of unintentionally creating more bomb makers for the Taliban. These classes built our understanding of Afghan culture.

"Win the hearts and minds," isn't limited to military operations. Politicians often use a similar tactic, attempting to connect personally with voters to build trust. In this analogy, the opposing political party plays the role of the Taliban (or at least that's what politicians are trying to convince you of). The problem in both the military and political scenarios is that these efforts seem more strategic than sincere.

The essence of true influence lies in authentic care for people—whether it's your team, your community, or an entire nation. It's about seeing individuals for more than their utility in a work environment or a strategic objective. Sometimes, my military leaders did not even bother to win the hearts and minds of their own unit.

November 5, 2009, I arrived in Afghanistan. Bagram Airfield was a hub of constant movement where units of soldiers would regularly pass through. The temporary billets were filled with fatigue and anticipation, housing hundreds of soldiers in a state of limbo as they awaited their orders to venture into the remote parts of Afghanistan.

Personal space and privacy were a luxury, so I found my temporary residence among the many rows of cots. In the air was the smell of sweat and diesel, due to the running generators and soldiers working in the unrelenting heat of the Afghan sun.

My cot was nestled amongst soldiers from a sister unit. They had arrived a week ahead of us but were destined for a different path. They were set to head south in a matter of hours, their gear packed.

The tension was high in the barracks. The master sergeant in charge of the other crew spoke up. His voice cut through the morning like a cold blade. They lounged on their cots until his sharp reprimand snapped them back to reality.

"You are all sitting on your ass when you need to be getting ready," the master sergeant barked, his voice echoing off the walls.

The soldiers scrambled to their feet. However, one specialist remained where he was, with evident fatigue in his eyes. He was met with the master sergeant's piercing gaze.

"Specialist Day, you obviously do not understand what the hell is going on. There will be repercussions if you do not have your shit ready in an hour," the master sergeant warned. His words were heavy with authority and an unspoken threat that lingered.

Specialist Day muttered under his breath after the sergeant walked away, "Asshole doesn't even know my first name, but he expects us to jump out of our ass for him."

The comment was low, but not low enough. Others heard and nodded in agreement. Specialist Day did not care.

He had seen how the master sergeant operated—how he demanded an unwavering commitment to his orders. Yet, he offered little in return. No camaraderie, no personal connection, just cold expectations, and the looming fear of "repercussion". It was a leadership of fear, not of inspiration.

Specialist Day sat on his cot, being led by a man who did not even bother to learn his name. To that master sergeant, Christopher Day was invisible.

How Well Do You Know Your Team?

How well you know your team members extends far beyond knowing their names. It's diving deep into their lives outside the office. Ask yourself: *How do they spend their downtime? Are they balancing family life? Or striving for personal goals like buying real estate or continuing their degree?* Maybe they're coaching their kids' sports events or taking care of an elderly relative. These insights are just the starting point in truly understanding the individuals in your team. Embracing these questions makes you a more empathetic and effective leader and creates stronger bonds among your team. Get out of the office and connect with your team on a human level.

Remember, it's not just about the information you gather, it's the connection you make. Like Ralph Waldo Emerson said, "It's not the destination, it's the journey." This journey of understanding is where you build respect.

Unsure how to start these conversations? Start with open-ended inquiries that delve into areas such as personal challenges, accomplishments, and meaningful experiences. These questions will not only kickstart the conversation but also pave the way for a deeper and more insightful dialogue.

What Could You Ask?

Are they married? If so, what are the names of their spouse and children? What sports do their kids play? Have they ever lived anywhere else? What was their favorite place and why? What is their favorite restaurant? Then go try it and let them know what you think. What is their favorite sports team? What is their favorite place to vacation? Do they play any instruments? What one would they choose to learn and why?

A Short Pencil

Our distinguished leader, Brigade Commander Colonel Bennett, made his way to our company one day, not for a grand inspection, but for a simple check-in. A brigade is a force of nearly 5,000 individuals, organized into three infantry battalions, each with just over 1,000 soldiers. Alongside them, there is a support battalion and the brigade staff. At that time, my company was roughly around 180 infantry soldiers.

Colonel Bennett's visits were different. They were not grand spectacles with soldiers in crisp uniforms and polished boots. There were no entourages of lieutenants and captains trailing behind him. He did not come to play "gotcha" or catch anyone in the wrong. He came to have real conversations with the soldiers. Usually, neither the company commander nor I knew about his visits until he was on his way out. These encounters could last anywhere from fifteen minutes to an hour. They could happen anywhere, anytime.

One day, Colonel Bennett engaged Private First Class Johnson in conversation. Johnson's wife, a fellow soldier in the brigade, was scheduled to deploy to Afghanistan in two weeks for a year-long assignment. Johnson expressed his nervousness, despite knowing they were thoroughly prepared. After a few questions, Colonel Bennett made a simple but significant statement: "Let's make sure the First Sergeant gives you time off before she goes."

As we were leaving the building together, Colonel Bennett stopped, pulled out a small notebook, and jotted down a quick note.

That note he scribbled down could've been about any number of vital things he had to keep track of. Maybe it was a note he needed to pass on to the commanding general. Or perhaps it was an order for the array of battalion commanders under his lead. It might even have been crucial information about the coming deployment to Afghanistan.

Instead, the note read, "Connect with Private First Class Johnson beginning of September. Wife / Alyssa deploys." He looked up at me and said, "A short pencil is better than a long memory."

This moment struck me. Here was a leader with an incredibly busy schedule, taking time to ensure a soldier's emotional well-being. A month later, true to his word, Colonel Bennett reappeared at our company, specifically asking for Johnson. He spent about five minutes talking to him, inquiring about his wife's send-off.

To Private First Class Johnson, this was the most significant and stressful part of his life at that time. Imagine the loyalty he felt for a colonel who not only remembered but also took time out of his hectic day to check in on him. That day Colonel Bennett gained not only Johnson's loyalty but also that of everyone around, including myself. This was the type of leader Colonel Bennett was.

I wonder how many notes filled Colonel Bennett's notebook during his visits with the other 4,800 troops under his command. It reminds me that, as leaders, sometimes the smallest gestures can have the most profound impact.

I walked with Colonel Bennett to the door. "Understand First Sergeant," he turned toward me, "this is the most important thing I will do today."

What is the most important thing you will do today?

We expect so much from our team. We ask them for their best effort daily. We ask them to care about the mission of the organization. We are asking them to take time away from their families. The least we could do is to get to know them.

Take a moment to understand your team, their struggles, and their needs. It is the personal connections that truly inspire loyalty and greatness. Your team members are not just your subordinates, they're your customers. Every day, you're running a unique kind of business where the product you offer is your leadership, and customer service is key.

Treat your team like they're walking into a five-star restaurant of leadership. Why? Because when your team feels valued, listened to, and respected, they're more likely to pass on that same top-notch service to the end customer.

Identify Your Customer

The traditional view of a business transaction involves three key elements: an employee, a product or service, and a customer.

Businesses generally expect employees to treat customers with such a high-quality service that it will result in referrals and repeats.

As a leader, you might think your role is to oversee the production and sale of goods. However, your true "product" is your leadership, and your primary "customer" is your team members. Your responsibility is to deliver leadership that encourages your team so they will welcome your guidance and mentorship.

For a healthcare company, the consumer is the doctor (not the patient). For a coffee shop, it's the barista (not the hipster). In the military, the soldiers are the customers (not the local population).

So, if you were to ask me, "Are you more concerned for your soldiers than the Afghan people?" My reply would be, "Yes, I am," and I make sure my soldiers know it. As a result, my soldiers will grasp the significance of our actions and the reasons behind them.

Providing effective leadership gives clear purpose, direction, and motivation, and inevitably leads to positive outcomes.

What Motivates Them?

Building a team requires an understanding of the motivations and backgrounds of each person. Understanding why they chose a career path sheds light on what influences their viewpoints and expectations.

Why did they join your team?

Their reasons could vary from pursuing education and training, financial necessities, a desire to travel the world, following a family tradition, seeking camaraderie, or the prestige of a title. In the military context, motivations might include a commitment to serve their nation, an idealized perception of combat, or an interest sparked by war-themed video games. Recognizing these motivations allows you to mentor and guide them toward their goals. What you think motivates a person and what actually motivates them might be very different.

To transform these individuals into a cohesive team willing to risk their lives for each other, it is essential to foster strong relationships. This process involves more than just relying on rank, it is about connecting with team members on a personal level, understanding their aspirations and fears, and helping them see the value and purpose in their actions beyond their initial reasons for joining.

You Don't Have to be Friends

In leadership, the lines between friendship and influence often blur—but they shouldn't. Effective leadership isn't about bonding over drinks or swapping weekend stories. It's realizing that knowing someone on a personal level doesn't require a personal friendship. Let's face it, some personalities just don't mix. But that can't stop you from leading. You have to learn to listen, because everyone, even the guy who can't stop talking about the dietary habits of his cats, may have a story worth hearing.

Adopt the stance of a mentor, not a drinking buddy. Treat everyone with the same level of fairness. Set boundaries and remember your mission isn't to win friends, it's to earn respect, trust, and maybe, just maybe, a little bit of admiration for not crossing into the "let's be buddies" territory.

The Difference of Two Leaders

The Chaplain

I recently had a conversation with an Air Force chaplain asking about what his job is like and how he dedicates his time. It was apparent his duties weren't limited to what you might expect. "So, aside from the traditional roles, what else do you find yourself doing?" I asked. He smiled, sharing stories of cooking pancakes for office staff, planning skiing trips, and even leading river floating excursions for team building. He finished by saying, "What I mainly do is get to know people."

The chaplain makes it a point to talk to everyone, asking about their lives, families, and work. It's this mindset that makes him so approachable. Interestingly, it seemed that his religious duties, or "spreading the word," often felt like a secondary aspect of his job. The chaplain's role, regardless of his own religious beliefs, involves catering to all faiths. Because of this approach, it's clear that folks feel comfortable coming to him with any kind of issue, personal or work-related.

The Counselor

There are also psychiatrists throughout the military. Just like in civilian settings, they play the role of counselor. They have responsibilities that occasionally overlap with the chaplain's. Both roles exist to offer emotional and psychological support to soldiers in need. Psychiatrists give counsel to soldiers struggling with post-traumatic stress disorder, high work demands, and even relationship issues. Their role is focused on addressing these problems and helping soldiers return to their duties. Their approach can feel more formal with less personal interaction. If you need to see them, schedule an appointment.

Although maintaining professional distance has its advantages, it can create a sense among many soldiers of psychiatrists not being approachable. The limited personal engagement may create a barrier for service members who might need support but feel reluctant to seek it due to the perceived formality in the relationship with these leaders. There is also a prevailing belief in these formal settings that expressing concerns could lead to professional consequences for the soldier. This includes fears of being removed from duty or facing challenges in career advancement.

Navigating the fine line between professional distance and approachability is key, and it really depends on the organization and the individuals within. You must empower and foster connections across your team, tailoring your approach to fit everyone's unique style. Leaders, take a hard look at how approachable your leadership really is. Ask yourself: *Do my team members feel at ease bringing their concerns to me?* Create an environment where everyone feels confident and comfortable speaking up, where there's a culture of openness and trust.

Recognize those on you team not just as subordinates, but as unique individuals with their own stories and drives. This approach will build a diverse group into a cohesive unit, united and ready to tackle challenges head-on. They'll stand behind you as their leader, and, more importantly, beside each other.

100 Stories

How is the bond so strong for people who have served together? Through hundreds of stories.

In the relentless grind of combat, days blur into months. You're with your squad, your brothers in arms, every hour of every day. Meals are shared in the few moments of calm. You sleep mere feet apart. You lay in the same dirt and often share the same blanket. You spend exhausted nights huddled behind a rock wall with a buddy, scanning for any sign of the enemy, so the rest of your team can catch a few minutes of sleep. With an ever-present danger, you come to know these people more deeply than you ever thought possible.

It's not just the laughs—though there are plenty, it's also the tears in moments of vulnerability when someone's guard slips, revealing the weight they carry. It's stories told during the long nights—tales of childhood, family, and dreams.

Hundreds of stories forge an unbreakable bond that time can't erode. These friendships will last a lifetime. It's in these unfiltered interactions that we find ourselves truly bound, not just by duty, but by something far deeper.

The bonds and loyalty between us are forged through human interaction, creating a mutual connection from leader to follower and from follower to leader.

KEY POINTS

- Know your team beyond their names—see them as more than just cogs in the machine.

- Build real connections—care about them as much as you expect them to care about the job.

- Be approachable—make sure everyone feels comfortable bringing their concerns to you.

- Follow up—make sure your team knows they're valued and heard.

- Get out of your office—connect with your people on a real level, learn their stories, and find common ground.

- Treat every day as a chance to focus on what matters—your team is your priority, treat them like it.

3

Check Your Ego

"My opponent is my teacher; my ego is my enemy."
– Renzo Gracie

After years of hard work, you've finally reached a pivotal moment in your career: leading your own team. With considerably more power and responsibility, you are now in a position to impact the daily lives of everyone under your guidance. Your team will look to you for answers, wisdom, advice, and mentoring. You will now have the opportunity to test your fieldcraft.

However, as Uncle Ben famously said to Peter Parker, "With great power comes great responsibility." Part of this responsibility is ensuring that your newfound power does not go to your head, a task easier said than done. This is the advice I always give to emerging leaders: "Your ego will destroy a team faster than you took charge of it." This also serves as an important reminder for everyone, regardless of their experience or position. In fact, the higher your rank, the more susceptible you become to the allure of its power.

To be an effective leader, it is crucial to understand the concept of ego. The definition of ego varies among great philosophers, as evidenced by the plethora of books on the topic. At its simplest, ego is how we see ourselves and how we want others to see us. However, in the context of leadership, ego becomes more complex. Being in a position of authority often feeds our ego, making it susceptible to unchecked growth. As we gain more power, our ego develops an increasing appetite for it. Elevated status also renders us more sensitive to criticism, leading to a fragile sense of self-worth.

It's important to recognize that everyone has an ego—it's an integral part of our identity. Our egos are shaped by our life experiences, and they significantly influence our beliefs and values.

Imagine yourself and your ego as two distinct individuals. The first one represents your true self—a rational thinker capable of seeing the big picture and identifying the necessary steps to achieve your goals. Your true self is confident and self-aware. The second individual represents your ego. Unlike your true self, your ego is emotionally driven and can often be quite immature. It exaggerates your self-importance and can become fragile if not managed properly.

The primary role of your ego is to protect you. Consider the ego as your personal bodyguard, always on high alert, especially when someone attacks your self-image.

Picture this: you pride yourself on being intelligent, but then someone dares to question your smarts. Your ego reacts emotionally, causing you to respond with all guns blazing, driving you to strike back, to prove them wrong. It's like a knee-jerk reaction to shield the image you've meticulously crafted.

Your ego is your hype-man, constantly amplifying whatever makes you shine brighter than the rest. It's all about staying on top, wielding power, and clinching victories. Its aim? To dazzle the crowd, to avoid even the slightest hint of embarrassment, failure, or appearing foolish. It's your ego's mission to keep your reputation spotless.

There are telltale signs when the ego takes the driver's seat. A classic one is hoarding information like it's wartime gold. Think about it—someone sits in on a top-level meeting, ears catching every detail. They become the know-it-all, the gatekeeper of information. They know details that others don't. Knowledge is power, and power feels phenomenal. It makes them feel important when they delve out tib bits of information.

But, for leaders, this approach is a death sentence for success. Information fuels motivation; the more people know, the more fired up they are to chip in. Keeping others in the dark is how you kill potential, and fast.

Some of the most common signs of an out-of-control ego are when you become overly reactive to feedback, criticism, or challenges to your authority. Instead of being open to constructive input about your idea, you may feel threatened or attacked, leading you to become defensive and lash out at others.

Back in 2005, in the dust and turmoil of northern Iraq, our company faced a persistent enemy. For a month, we were attacked almost daily while trying to keep a vital supply route clear. Captain Rivers, our company commander, spearheaded a strategy with his lieutenants. The plan? Occupy a building along Route Volkswagen, hoping to thwart enemy ambushes on US convoys. It seemed sound on paper—we'd gain visibility over the trouble area. During the final briefing, as the map unfolded before us, a glaring truth struck me. We were about to paint ourselves into a corner, with no easy out. In the overwatch position, our visibility would be excellent, but we lacked any view of our own access and egress routes. Captain Rivers, whom I respected greatly, had overlooked this significant detail. The enemy could easily adapt, striking our only paths in and out. A fatal flaw.

During the briefing, I couldn't hold back. What seemed obvious to me was a blind spot in the plan. I proposed an alternative location—one that offered better visibility of our approach and retreat routes. My suggestion was a bit of a pivot, but it put Captain Rivers in a tight spot, under the scrutinizing gaze of twenty subordinates.

Despite my gut screaming caution, his response was firm: "Our current plan stands." It was a tough pill to swallow. The moment should have been a course on the importance of fluidity in leadership. War, as we learned, requires fluidity.

For two relentless weeks, we held that building, bracing against challenges more difficult than we expected. The IEDs didn't waste time; day two. Boom! Boom! Two earth-shaking blasts jolted us back, but we were lucky; no lives lost.

By day seven, the story took a turn. Two more IEDs lay in wait on the eastern approach. This time with less luck. Two of ours were hit, evacuated with shrapnel wounds, and heads rattling from concussions.

Then came day ten, etched in my memory like a scar. A white pillowcase—or so it seemed—bulged in the open. A trap baited and waiting. Our explosive ordinance team inched toward it, stopping at what should have been a safe distance. But that's when the earth beneath them ripped apart. The real IED was buried not where we expected it but precisely where they knew we'd be. The pillowcase was just a decoy taunting our predictability.

The aftermath of the blast was severe. Two of our best explosive techs lay critically wounded, their lives hanging by a thread. Another soldier, caught in the explosion, was also down.

It was clear this rate of loss was a no-go. Something had to give, and fast. We had to adapt, or risk losing more than just our supply routes.

In the aftermath, questions haunted me. Was this a case of a commander's ego overriding his tactical judgment? Did my direct challenge to his authority in front of the team influence his decision? Could I have presented my concerns differently for a better outcome?

Feeling defensive when directly challenged is one indicator of a large ego. Another clue is constantly seeking recognition and validation, a need for applause and nods of approval. A big ego craves that spotlight even if it means shoving others into the shadows. You may also feel a sense of A third indicator of a big ego is a sense of entitlement or becoming overly competitive with others. As if you're in a race with your elbows out, ready to outrun anyone else because the winner's podium was meant for only you. You believe you deserve success and recognition more than anyone else.

Lacking empathy or concern for others also indicates a harmful ego. If your ego is in control, you may focus more on your needs and desires than those around you. You may also struggle to see things from another person's perspective or dismiss their concerns and feelings. Finally, being inflexible and unwilling to change your approach or adapt to new situations can signal that your ego is running the show. Suppose you are more concerned with maintaining control or being right than with finding the best solution. In that case, you may be allowing your ego to hold you back from success.

Signs of an Unhealthy Ego:

- Believing you are better than others and needing to feel in charge or relevant.

- Feeling you deserve more than others and that certain tasks are beneath you.

- Rejecting ideas that aren't your own.

- Believing your title is a key component of your identity.

- Dislike being corrected, especially in public.

- Seeing competition as an opportunity to show you are better than others.

- Feeling envious, jealous, and angry when others are successful.

- Avoiding weaknesses and redirecting the focus to your strengths.

- Viewing failures as setbacks and often blame others.

- Protecting yourself from shame or failure.

- Seeing others' success as hindering your own.

- Interrupting others often.

- Lacking selflessness and constantly comparing yourself to others.

- Needing and valuing external approval and validation.

- Having obsessive thoughts and wanting to control what others think about you.

Signs of a Healthy Ego:

- Asking for help.

- Setting boundaries.

- Acknowledging and appreciating others' input.

- Having an abundance mindset.

- Celebrating the success of others.

- Listening with an open mind.

- Being confident without needing to tell everyone how capable you are.

- Not needing to prove your intelligence.

- Understanding that perfection does not exist.

- Embracing your flaws and looking for opportunities to grow.

- Not changing who you are to fit in.

- Competing against yourself in the pursuit of self-improvement.

- Not requiring external validation.

Egotistical or Confident

Major Benton had a great resume. He had done "cool-guy shit" most people don't get to do. He spent time working with and supporting government three-letter agencies. He worked in parts of the world most of us hadn't heard of.

How did we all know this? He would mention it every chance he had and would look for any opportunity to work it into a conversation. He would mention it to the point of exhaustion. You would think, given his background, he wouldn't feel the need to impress others. Major Benton acted as though his background made him above the standards the rest of us followed.

He put so much energy into getting his subordinates to think he was a badass, he lacked awareness of how people really felt. After all, his resume was great, and he had tremendous experience. But because of his constant need to impress, he established a reputation as a braggart. It destroyed his credibility throughout the organization, especially amongst the lower-ranking members he constantly tried to impress. Even when he had valuable input, it was very difficult to get past his reputation and take him seriously. His ego's need to impress destroyed his credibility.

There is a lot of confusion about the difference between being egotistical and confident. I believe confidence is one of the top characteristics of a leader. It's essential for any leader to display confidence in order for their team to have faith in them. It is very frustrating if you have tried to work for someone who is unsure of themselves. They are reluctant to make a decision because they fear being wrong. They constantly seek the approval of everyone around them by overcompensating in their behaviors, such as trying to get people to like them. A leader that lacks confidence in combat will get people killed. We often say in the military that a wrong decision is better than no decision. If you make any decision, at least you are moving forward and trying to improve your position. If you don't make a decision, you will remain in the same place you started while the enemy maneuvers to gain the advantage. We say this to give new leaders the confidence to make decisions. We want to make it okay to be wrong and learn from it. Eventually, these experiences lead to more and more confidence.

Experience and education instill confidence in a leader. You are confident because you know what needs to be done or that you have the capacity to figure it out. I believe that is where my confidence comes from. I know that even if I'm not sure how to do something, I have an excellent track record of being able to figure it out. Consulting with my team is one way of keeping things together. Between their experience and mine, we can figure it out. Even if the "doing" process becomes trial and error. My confidence comes from doing and seeing things incorrectly enough times that I learned from it. Now, I know what I'm capable of accomplishing.

So how is ego different from confidence? Determine if a person is simply proud of their accomplishments or if they actually believe they are better than everyone. Arrogance is rooted in insecurity and is a form of egotism. It's the need to impress on others your superiority. Although we want a leader with confidence, we will avoid interacting with arrogant people. How good of a leader can you be if your team can't stand to be around you?

Tony Stark had a piece of advice for Peter Parker: "If you're nothing without the suit then you shouldn't have it." You don't deserve to wear that rank if you are nothing without your rank or job title. Your character, work ethic, trustworthiness, and competence make you deserve the rank. Without those things, the rank means nothing. In fact, wearing it discredits the rank and others who wear it. When you act in an unbecoming and undeserving way, it cheapens the rank.

Influence vs. Manipulation

A leader will influence individuals or his team toward a mutually beneficial goal, while manipulation makes people commit to a goal that benefits few (often to the detriment of others). Manipulative tactics are prevalent in politics and media, where individuals selectively present biased perspectives and cherry-picked facts to influence others into agreeing with their viewpoints. Such behavior can stem from an inflated ego and a desire to assert dominance.

In contrast, true influence is wielded by presenting all relevant information, weighing the pros and cons of various options, and charting a course of action that is most likely to yield positive results. We are all susceptible to manipulation due to our innate tendency to seek agreement from others. However, we can counteract this tendency by adopting an objective, introspective approach to problem-solving. We must remain mindful of whether or not we inadvertently promote one-sided perspectives and close ourselves off to alternative solutions due to our ego. Only then can we lead with integrity and ensure our actions are guided by a genuine desire to serve the greater good.

I'm always fascinated to see how often individuals respond and communicate with their egos. By actively listening to conversations around us, we can gain insight into why people say what they do. Some individuals may make subtle comments to assert their importance, control, or likability. For instance, they may seek admiration through comments about their possessions or prestigious neighborhood. However, such behavior often stems from a sense of insecurity regarding how others perceive them. While mentioning their car doesn't always mean someone is trying to impress others, it becomes clear over time whether their intention is to impress or simply to make conversation. Conversations with such individuals can often leave us with a sense of their shallow character, as they lack self-awareness and the ability to perceive our awareness.

Our egos may feel insulted, as if the other person assumes we lack the intelligence to see through their facade. Nonetheless, by maintaining an awareness, we can find humor in these comments rather than feeling insulted. Ultimately, we can develop more meaningful connections with others by cultivating greater self-awareness and learning to communicate with integrity and authenticity.

Keeping Your Ego in Check – Respond Don't React

This is how we can counter the ego's desire to remain in control. The secret lies in four simple steps.

Step 1: Awareness.

Remember, everyone has an ego. Its function is to protect you. Be aware of how the comments of others make you feel. For example, if someone disagrees with you, does it make you defensive? If a belief you hold to be true is questioned, do you tend to attack the other person, even if it's just in your head? If you are corrected, do you talk shit about them in your head, or do you welcome feedback?

It's good practice to pay attention to others and their interactions. Notice when they are responding with emotion. Notice the subtle comments made to make themselves feel big or make others look small.

Step 2: Pause before you respond.

Practice a pause and take the extra moment to decide if your instinct is to respond from the ego. Are you only saying this because you feel attacked or challenged?

Step 3: Don't take it personally.

Let them have their reaction and don't take it personally. Try to see their perspective on the situation. Do they have insight or knowledge that you don't? It could be their comments and actions have nothing to do with you and everything to do with them. Is it their poor self-esteem that drives their egotistical response?

Step 4: Keep your desired outcome in mind.

Before taking any action, determine what response will bring you closer to your goal or push you further away. If your goal involves building a team, then cutting the other person down because of your own insecurities is unlikely to help you achieve it. Your team already knows you're in charge, so there's no need to constantly remind them. Instead, acknowledge their contributions and avoid attacking them.

To achieve your goal, you need to communicate why your chosen outcome is beneficial for the team and how you plan to get there. By influencing their decision-making and thought processes, you can help them understand the value of your goal and how it aligns with their own objectives. However, if your goal is driven solely by your own ego and benefits only you or a small group of people, then you may find yourself resorting to manipulation rather than influence. You'll present one-sided facts and opinions that support your desired outcome rather than taking a more balanced approach that considers the needs and perspectives of all stakeholders involved. In short, the key to achieving your goal lies in building a solid and motivated team rather than tearing them down.

Ego Signals Immaturity

The people who are impressed by your egotistical comments are generally immature. It's entertaining to watch two egotistical people have a conversation. Neither listens to what the other is saying. They are simply waiting for an opportunity to interject and talk about themselves with little awareness. Often interrupting the other person, they feel what they have to say is more important. They don't listen, and it becomes a "one-up" contest of who has the more significant story. It becomes a battle of egos, which is quite entertaining. It's like a dueling piano bar and arguably just as fun. We have all witnessed this, and hopefully, this isn't you.

Look for opportunities to keep your mouth shut even when your ego wants to speak up. Identify when your ego wants to respond rather than your logical self. Know that your emotional response is generally unproductive and will often lead to a battle of egos. Also, be aware of your ego talking inside your head, feeding you bullshit. It's essential to break that chain of thought so you can be more logical in your decision-making. Instead of trying to put someone in their place and attacking their ego, focus on influencing them towards the best outcome. Start by acknowledging their efforts and contributions and then guide them towards a solution that benefits everyone. For example, you could say, "I appreciate everyone's hard work and input on this. Let's focus on how we can work together to achieve the best possible outcome." By approaching the situation this way, you can avoid damaging relationships and achieve a more positive outcome for everyone involved.

There will be times when another person's ego questions your ability or authority. That person's need to feel in charge and noticed might make you feel vulnerable. He wants to feel important. His ego's emotional response is to make the authority and competence of others questionable. Have the maturity to use the four steps mentioned with the primary objective of what's going to move your team closer to the goal line. A battle of egos surely won't.

Ego: A Hurdle to Effective Leadership

Ego can pose a major obstacle to effective leadership. When leaders are motivated primarily by ego, they prioritize their own interests and desires over those of their team. This can create an environment where collaboration is difficult, feedback is unwelcome, and admitting mistakes is taboo. Leaders may also attempt to control or manipulate their team, instead of fostering an environment of trust and open communication. This can make it difficult for leaders to make decisions that benefit the team or organization. Ultimately, the most successful leaders are those who can balance their own ego with a deep commitment to the success and well-being of their team. A balanced ego can be difficult to keep in check, especially if it has run amuck for some time. Remember that fieldcraft takes time and practice, and in this case, self-awareness.

KEY POINTS

- Check yourself. Know when your ego is calling the shots and get real about what's healthy and what's not.

- Pause before you react—don't let emotions or competition control you.

- Listen with an open mind and try to see things from the other side.

- Be confident without needing applause—don't hoard info or seek validation.

- Stay sharp—know how ego shows up in yourself and others, and don't take things personally.

- Learn to shut up when your ego wants to speak—use your head, not your pride.

- Trust yourself to know what needs to get done—or figure it out when you don't.

4
Commander's Intent

"A lack of communication leaves fear and doubt."
– KELLAN LUTZ

MILITARY LEADERSHIP IS NO EASY FEAT. It demands a blend of confidence and control—qualities that represent physical and mental resilience. Yet, one skill stands out the most: Clear and effective communication. This skill is the linchpin of successful leadership. You may have tactical brilliance, but if your ideas are not communicated clearly, your strategy becomes useless. Communication is especially important amid the confusion of combat. The ever-present stress, including the possibility of being killed, makes people behave unpredictably. Thus, it is crucially important to ensure your instruction is clear.

Two of the military's most powerful communication models are Commander's Intent and decentralized command.

These core concepts are not unique to the military, they also hold significant value in the civilian sector. Understanding and implementing these concepts can greatly enhance leadership and communication skills, making teams more informed and knowledgeable. More importantly, these models are adaptable, providing reassurance and flexibility in various situations.

Commander's Intent

Commander's Intent is best defined as a single declarative statement that communicates the ultimate result or goal of the commander, CEO, manager, or team leader, to all team members. It gives a crystal-clear understanding of the desired outcome and is designed to be ingrained in everyone's memory. The Commander's Intent can be as short as a single sentence. The clarity of this statement provides a sense of security and focus, even in the midst of confusion and stress. Again, when survival is on the line, human behavior shifts dramatically. Brevity resonates more clearly with a confused or stressed mind.

Decentralized Command

Decentralized command is the leadership practice of allowing people the autonomy to make decisions and take action without direct supervision. It involves managing teams or individuals who are distributed in a manner that makes direct supervision impossible, whether due to physical distance, geographical separation, or the nature of their duties. This is particularly important in modern combat where the distance between senior leaders and those executing the mission can be huge.

Without a clear Commander's Intent, decentralized command is not possible. However, when the desired outcome is clear, individuals are empowered to make sound decisions independently.

In the military, the Commander's Intent spells out what success looks like for a mission. For example, "Seize and control the enemy compound in (Specific Location) Village". With that intent laid out, everyone on the team knows the endgame and can make moves on their own when needed. This clarity is a lifeline when the shit hits the fan.

This concept also applies in the civilian world. Every organization, store, branch, or team should operate with a Commander's Intent—call it the boss's intent if need be—a definitive statement that remains top of mind for every team member throughout the day. This intent serves as a unifying force, just like in military operations, supporting the organization's mission and aligning everyone towards a common goal.

Have you ever been to a restaurant and had to return your food for some reason, only to be met with an argumentative reaction, or one of annoyance or irritation at having to remake or change the order? At Starbucks, there is a clear Commander's Intent; Make sure the customer is happy with their beverage every time. If you were to return your drink at Starbucks, the baristas, almost unanimously, will happily and immediately remake your beverage to ensure you are satisfied. Every single staff member has the authority to "make it right" no matter the complaint and is encouraged to do so with positivity. This is consistent across stores, whether you're in Oregon or New York. Starbucks has done a good job of ensuring their employees clearly understand the mission.

Consider your own organization; What do you believe is the most important thing your team needs to remember when conducting daily tasks?

The Military Operation's Order (OPORD)

In the military, mission planning is methodical, precise, and structured. We use a massive checklist to ensure we've covered every angle and to cut the risk of oversight to as close to zero as possible. The result of this checklist is a military operation's order (OPORD), a written page comprised of five sections that is the guide used for any event or mission. It concisely states the details of what needs to happen and what the end state needs to be.

For reference, the five-section format of a military OPORD includes situation, mission statement, execution, service and support, and communication (command and signal). We are homing in on just two as they are especially crucial for the scope of this chapter: the mission statement and execution sections.

The first part of the operations order I want to discuss is the **Mission Statement**. This statement is straightforward, answering the questions of who, what, when, where, and why for the operation. It provides an overview that prevents needless complexity.

The next is the more elaborate **Execution** portion. It outlines the mission goal and gives specific directives for each element. The execution paragraph begins with the Commander's Intent. If the mission statement outlines what you're going to do, then the Commander's Intent statement defines the end result, when that mission is considered successful.

A Clear Mission

Despite a complicated planning process, the Commander's Intent statement is short and simple, declaring the desired outcome and affording us an overarching understanding of the entire mission. It is the mission's beacon—a target etched into everyone's consciousness.

In an OPORD, every unit will be assigned a different role in the mission. The leader of each unit knows what their team's responsibilities are and communicates a Commander's Intent for that unit. This supports the higher commander's intent, which every individual in the unit is also aware of.

For instance, if a battalion commander's mission involves raiding an enemy training compound, every company within the battalion assumes specific responsibilities that contribute to the overarching objective. The battalion commander's intent could be seizing, controlling, and denying the enemy access to the compound. The first company would be assigned to secure the perimeter, thwarting reinforcements, and preventing enemy escape. The second company would breach the compound and establish a foothold, while the third would complete a comprehensive building-by-building clearing. Despite their distinct functions, all companies align with the battalion's primary aim: seizing and controlling the compound. And all parties involved—every soldier in every company—will know their commander's intent for their specific task, as well as the overarching intent of the mission. And when everything goes to hell, even if the leaders are taken out, the soldiers can carry on—no matter what, because they know exactly what the mission demands, no questions asked.

In the movie *Saving Private Ryan*, Captain Miller (played by Tom Hanks) embarks on a mission across the Western Front during WWII to locate the last surviving son of Mrs. Margaret Ryan. Private Ryan's three brothers were killed in action, prompting the US Government to intervene, ensuring that the Ryan family would not lose all four sons in the war. In contrast to my combat experiences, Captain Miller faced limited information about his mission and was simply armed with a clear directive to find Ryan and bring him home. "Find Private Ryan" was the Commander's Intent, empowering Captain Miller to navigate the battlefield, making life and death decisions while remaining steadfastly aligned with the mission's ultimate objective.

Grasping the intent of the operation is a pivotal make-or-break factor. Circumstances can spiral into chaos, just like they did during Captain Miller's search. However, a team aligned with the Commander's Intent can adapt, making decisions that support the overarching objective despite evolving scenarios, because they understand what the ultimate objective is. Remember, the enemy also shapes outcomes. It's plausible to find yourself amidst an operation gone to shit. The command structure is gone, communication devices falter, and worst-case scenario, some decision-makers are killed. Despite this mayhem, one unifying factor endures, the Commander's Intent. If there is only one man left standing, he will know how to make decisions and adjust to the current circumstances to fulfill the goal of the mission. This statement forges unity amidst the chaos.

Mission Statement versus Commander's Intent

Don't confuse a company's mission statement with Commander's Intent. In the civilian world, too often, a company's mission statement is just fluff—something glanced at on a website or printed on a poster in a lobby that sounds good to customers. Though occasionally you will find companies with clear declarative mission statements.

What's your company's mission, and can you, as a leader, step up to support it? If your team doesn't have a clear goal, it's on you to define one. As a leader, your job is to forge an intent that's locked in with the company's mission and vision. Distribute this clear statement to all team members and ensure they know it. Without your direct oversight, this intent needs to be the compass for every decision your team makes. What is the one principle you want your team to have etched in their minds when they're in the trenches making calls? It's not the flowery mission statement, it's the commander's intent. Make it short, sharp, and make it memorable.

Loosen the Reins

In the realm of team leadership, inexperienced leaders often keep a very tight rein on their teams, driven by an urge to exert control over chaotic scenarios. At its root, this is a fear of failure. In the chaos, inexperienced leaders are breathing down the necks of their team, constantly seeking updates, demanding explanations, and shadowing their every move. These leaders stay physically close, offering whatever support or instruction they feel is needed, micromanaging the situation. Sound familiar? We've all witnessed those ceaseless radio transmissions or in more familiar situations, relentless phone calls and emails from superiors who want to stay "in the loop". What unfolds among your team members is a sense of distrust, as if their abilities are under constant doubt and their actions are perpetually under a microscope. Is this level of scrutiny necessary? While every leader has their unique style, it's crucial to allow room for personal development and self-sufficiency.

As a civilian business expands, the need for decentralized command naturally grows. In smaller businesses, it's common for leaders to juggle multiple roles, handling everything from sales and marketing to accounting and beyond. However, as the company grows and new employees are brought on board, it can be challenging for leaders to relinquish tasks they've been accustomed to managing themselves and trust that others will get the job done.

The Fog of War

From our position, Captain Lewis peered into the dense fog, unable to see the movement of either platoon. There was a blurred line between our front line and our enemies.

The fog of war, both literal and figurative, was a familiar adversary—one that hid friend from foe and certainty from doubt. Captain Lewis' inclination was to radio his platoons for an update, but his experience held him back. His expression urged us to be patient.

Our mission was to secure a region notorious for Taliban activity, particularly where they had been ambushing resupply convoys entering the valley. Intelligence reports from the previous evening indicated their presence. They were observed establishing positions on the elevated terrain overlooking the primary supply route.

"First Sergeant, are you able to get eyes on Walters' platoon yet?" Lewis asked.

"Negative, but they should be cresting over the high ground any minute," I replied.

We could hear the distant, muffled sounds of movement. Was it our men advancing or was it the enemy creeping closer?

We raised our rifles in the direction of the sound, ready to fire on the figures as they got closer. Holding steady, we waited until the last second when they were almost on top of us trying to identify who was approaching.

We hadn't been able to reach Lieutenant Walters on the radio all morning.

"Hold your fire," the command echoed through the lines.

"I can see them. It's second platoon," called out a soldier on the right flank.

Lieutenant Walters, face smeared with dirt and sweat, emerged from the mist. "Sir, we pushed all the way to the bridge," he said with confidence. "No sign of Taliban. We must've scared them off." The commander looked at him, at the map, then back to the lieutenant with an approving nod. "Well done, lieutenant."

A genuine leader possesses the capacity to step back, demonstrate patience, and allow their team to function with as much autonomy as possible. They accept the fog of war, being in the eye of the storm without falling victim to it. They don't smother their team with an overbearing presence. Experience and maturity can empower a good leader to grasp the big picture despite the chaos. Trust in your team, trust from your team, and good training are all necessary for successful decentralized command of empowered individuals.

How is this level of comprehensive situational awareness achieved? It hinges on practice and continuous education until it becomes second nature. You must rehearse embracing the chaos and resisting the urge to enclose everything within your grasp. Pause and evaluate the entire situation to determine the best course of action.

In the spring of 2004, I found myself in a small town approximately fifteen kilometers to the north of Tall Afar, Iraq. Near the Syrian border, there was a known pit stop frequently utilized by anti-American combatants en route to Iraq to have their chance at battle. It was recognized as a strategic location due to its significance in accommodating these fighters.

Our platoon was tasked with securing an Iraqi police station that had fallen under enemy control the day before. Our assignment was securing and maintaining control over the station until Iraqi reinforcements could be deployed to the location. Intelligence reports available at that time indicated that the adversaries had successfully captured the station, looted its contents, and vacated. The enemy was notorious for their practice of leaving explosive devices rigged to tripwires in situations just like this.

Potential American reinforcements were estimated to be three hours away, even under the most favorable circumstances. We found ourselves in relative isolation within a town about which we knew very little. Despite being hours away, our commander exhibited confidence in our capability to make strategic decisions. It was a moment where an excess of directives could have hindered our ability to adapt to the evolving circumstances. Flexibility was necessary, and our commander recognized this.

Our platoon made numerous tactical decisions within the forty-eight-hour timeframe, each contributing to the ultimate success. During this critical period, the luxury of seeking guidance was nonexistent, leaving us alone to navigate the challenges.

We were faced with the decision of whether to enter the compound without the support of an explosive ordinance team. We would have to rely on the training we had received and use a high level of situational awareness. We entered the compound to discover the enemy had begun to place one of the biggest improvised explosives I had ever seen. Luckily, our arrival prevented them from putting the final touches on the detonation device.

Considering the distance from reinforcements, we immediately chose to clear two areas for a medical helicopter to land in case we started to take casualties. One was nearby, and the alternate was a kilometer north of town. We ended up using it for resupplying our ammunition the following day.

Throughout the night, we deployed small teams beyond the immediate safety of the compound to gather intelligence on potential threats. We meticulously balanced the possible risks against the potential benefits. This strategy was to prevent the enemy from closing in on our position too easily and gain a solid understanding of our defensive setup. It also conveyed a clear message that we weren't merely present—we were fully prepared to engage in combat if necessary.

The following morning arrived. The sun cast its early morning glow, illuminating the surroundings. I cautiously peered over the cinderblock wall that stood three feet high, encircling the building's top. A glint caught my eye from the bushes across the street.

Keeping my voice low to avoid detection, I whispered into the radio, "I see something reflective. Can you see it, Sergeant Buchheim?" The response came quickly from my Alpha team leader, who was positioned elsewhere. "I see it too. Looks like two males, probably in their early 20s, observing us with binoculars. They're also on the phone."

Alarm bells rang in my head. Who were they communicating with? Could they be relaying crucial information about our position to potential adversaries? Realizing the urgency, I said, "They're dangerously close, less than 100 meters. We need to intercept them before they spill too much."

I reached out to the lieutenant, offering to lead a small team for the task while the rest of the platoon kept a vigilant watch. Without wasting a moment, I instructed, "Sergeant Buch, assemble three men. We need to move, now."

We advanced cautiously across the street, opting to circle around our position from behind to maintain the element of surprise. I radioed to our overwatch team, "Apache 2-3, this is Apache 2-2. Monitor the street and alert us if you notice any movement." A crisp response came back, "Roger, 2-2. We've got your back."

As we neared the bushes where we'd spotted the two males earlier, they suddenly emerged and sprinted down a nearby alley. Noticing they were unarmed, we held our fire. "Klein, swing around that corner and track their movement. Let's see where they're headed," I commanded. My team swiftly increased their speed.

Rounding the bend, Specialist Klein reported, "They've taken cover in another alley, about 20 meters ahead." I quickly checked our formation, ensuring we remained within the protective gaze of our overwatch. "Apache 2-6, this is Apache 2-2. We're advancing down the alleyway." The reassuring voice of 2-6 replied, "Understood, 2-2. We're monitoring the main road."

Advancing cautiously down the alley, we were about to round the corner when a sudden warning came through the radio. "2-2, something's off. The streets are deserted. Everyone's vanished." It was unusual for the streets to be empty at this hour.

Without warning, a deafening bang resonated, followed by a swooshing sound. A rocket propelled grenade had been fired, exploding just a short distance away, sending shockwaves and debris flying. We instinctively took cover against a nearby wall as bullets whizzed past, ricocheting off bricks and kicking up dust from the ground. The realization hit hard—we'd been lured into an ambush.

We retaliated with fierce determination. Our position was vulnerable, but our attackers had underestimated our firepower. Our overwatch team provided invaluable support, raining down bullets on the assailants. I signaled to Sergeant Buch, "We need a grenade over that wall, where the RPG came from!" As Buch reached for a grenade, another resounding boom echoed, followed by the familiar swoosh. The projectile was headed straight for us. In a twist of fate, it struck a pole in its path, diverting its trajectory, and crashing into a wall behind us.

Wasting no time, we lobbed our grenades, their subsequent explosions dampening the enemy's assault. With a renewed sense of urgency, we tactically retreated, making our way back to the compound. As we maneuvered, vehicles armed with powerful .50 caliber machine guns provided cover, relentlessly targeting, and suppressing the enemy's positions.

Upon safely returning to the compound, we took a moment to regroup and strategize. Analyzing the enemy's position, we decided to employ a tactic matching the one they had just used against us. The plan was to use the movement and noise of our vehicles as a diversion, drawing their attention away from our main force, which would then navigate a back alley unseen.

With the plan set in motion, the rumble of our vehicles filled the air, creating the perfect distraction. Meanwhile, our team stealthily advanced down the alley, reaching the building the ambush had been launched from. Bursting in with force and determination, we expected to confront the enemy, but what we found instead was an eerie silence. Signs of a hasty departure were evident: bandages, spent ammunition, and makeshift medical supplies littered the area. It was clear that the enemy had suffered casualties during our counterassault, but they had managed to regroup and evacuate their wounded, leaving behind an empty battlefield and lingering questions.

We were lucky that no one was lost that day. But the ambush taught us a tough lesson about eagerness and the patience to let a situation develop.

Indeed, even after falling for the ambush, our ability to quickly adapt and pivot was key. Our commander's trust in us to make on-the-spot decisions proved invaluable. We operated independently and developed a counter plan without delay. We understood the Commander's Intent, and the effectiveness of decentralized command was a success as a result.

By day four, reinforcements had rolled in. A beefed-up Iraqi police force, better equipped and ready to take on the opposition, was now on the ground. Despite nearly suffering disaster, my platoon's mission was a success.

5 Factors for Effective Decentralized Command

There are five primary factors that are pivotal for the effectiveness of a Commander's Intent and decentralized command:

1. Ensure understanding of the desired outcome.

Effective communication prioritizes quality over quantity. An effective leader selects words carefully, ensuring every word carries weight. Your team should not have to resort to mind reading to understand objectives. In the absence of orders, they must have the autonomy to make decisions aligned with a clear Commander's Intent.

2. Confirm your team is trained and ready.

To have confidence in your team's ability to accomplish the mission, you must equip them with training in all the essential skills required for success. This responsibility falls on your shoulders as a leader. Any deficiency in their training reflects a lapse in your leadership. Training is the glue that keeps soldiers together amidst the chaos.

3. Place trust in your team.

Proximity isn't necessary and distance often facilitates optimal performance. It's important to have faith in your team and their skills. Project your vision and bestow autonomy. Leaders who are accustomed to being intricately involved in every aspect of the mission may find it challenging to let go. However, there comes a point where a leaders' capacity becomes limited, and the team's performance hinges on the leaders' ability to let go.

4. Ensure communication is clear and ongoing.

You must consistently provide clarity, especially amidst chaos and confusion. Can you express your intent clearly or do you risk overwhelming your team with complexities thereby contributing to the confusion? There is such a thing as over-communicating, and I see it often. Excessive communication conveys to your team that you lack confidence in their competence or capacity to make decisions. People talk more when they are uncertain or scared, and war is the epitome of uncertainty.

5. Foster an environment of mutual accountability and feedback.

If you aren't going to set an example and hold yourself accountable for your mistakes, why should anyone else? Leaders must exemplify receptiveness to feedback, setting the tone for an environment where ideas are exchanged openly. Your team needs to have confidence that you'll execute your commitments and provide support when necessary. More on this in chapter nine.

The concept of the Commander's Intent extends beyond the military realm. Like the military, corporate communication should be direct and simple, ensuring comprehension and retention.

Every organization should establish an overarching intent beginning at the highest level. What fundamental idea should every team member be thinking about every day? Unlike in the military, civilian missions tend to be continuous and ongoing. They lack a single culminating event, an "enemy compound secured" moment. Rather, the focus might be on delivering optimal product quality and superior customer service. Having a clear intent, whether military or civilian, is key to powering through and completing the mission.

KEY POINTS

- Keep the mission clear and to the point—brevity hits harder and resonates.

- Lay out what matters most—then trust your team to get it done.

- Empower your team to make decisions—use decentralized command to keep things moving.

- Embrace failure whenever possible—it's how you build future leaders.

5

Delegate to the Lowest Level

"The best executive is the one who has sense enough to pick good people to do what he wants done, and self-restraint to keep from meddling with them while they do it."
– THEODORE ROOSEVELT

LEADERS ARE OVERBURDENED with stress and work that could easily be handed over to someone else. Our brains and bodies are not made for multitasking, yet most leaders choose to keep multiple tabs open in their heads. Multitasking will not only slow down your overall progress but also harm the quality of your work. This problem can be mitigated through delegation. What tasks, projects, decisions, or functions are you doing or working on right now that a subordinate could do? Delegating is shifting the authority to another person so they can complete a task or project. This does not mean that you entirely free yourself from the task. In the military, we say, "You can delegate authority but not responsibility." The last thing you want is for your team members to feel undervalued or unappreciated. Understand that every individual possesses unique skills, perspectives, and contributions that are vital to your collective success. Failing to acknowledge and recognize their value can lead to demotivation, decreased productivity, and ultimately hinder the growth and effectiveness of your team. Strive to create an environment where each person feels seen, heard, and valued for their unique abilities and ideas. By fostering a culture of appreciation and recognition, you ensure that your team members feel empowered and motivated to go above and beyond.

Create high expectations for your team: active engagement and proactive responses. Avoid making your team members feel like they are merely another cog in the wheel, performing mindless tasks. As individuals crave more creativity in the workplace, it is crucial for you to recognize and appreciate the unique contributions they bring to the table. This acknowledgment comes through valuing your team as an asset and recognizing each member as a leader in their own right.

It's Your Turn - Take the Mic

The day was fading, with the sun dipping closer to the horizon. Our afternoon had been consumed by airborne training operations. Executing parachuting drills from Blackhawk helicopters onto the sprawling drop zone, each six-man team taking its turn in a meticulously orchestrated sequence. Two helicopters constantly rotated to sustain the momentum. As teams landed, they discarded their kit bags, each containing a recently utilized parachute, then began the task of rigging for the next jump.

My rank was specialist at the time—just under three years in the service. I found myself assisting Staff Sergeant English in managing the drop zone. Staff Sergeant English adeptly coordinated the movement of the helicopters via radio transmission, ensuring their rotations were synchronized within the same airspace. His expertise was essential in maintaining a seamless flow of operations. My role focused on supporting his directives.

Staff Sergeant English handed me the radio. "Staley, take the mic. Keep an eye on Sidewinder and make sure he stays south of the 36 gridline while Strike is in the air," he instructed, emphasizing the importance of managing the helicopters' positions to avoid any airspace conflicts.

"Roger that, Sergeant," I replied. Taking the radio, I felt the weight of the task. My focus was now on ensuring Sidewinder remained in its designated airspace, to maintain a safe operating environment while Strike was conducting maneuvers. Careful coordination was imperative for the success of the night operations and the safety of all involved.

I managed the radio talking to both aircraft as Staff Sergeant English gave me guidance on what to look and listen for and say.

It really was impressive. His directions became fewer and fewer as I began to grasp the nuances of controlling the air traffic.

"Reaper 11, this is Strike 23, we are heading off station to refuel and will return at twenty hundred hours," came the radio communication from Strike 23, informing me of their imminent departure.

"Roger Strike 23, that is a good copy," I replied, acknowledging the message. With Strike 23 off station, it was clear Sidewinder would have the airspace to conduct one more drop before it would also need to refuel.

As I looked up, six parachutes blossomed in the dimming sky, a testament to the ongoing success of our operations. Just then, the radio crackled to life again. "Reaper 11, this is Sidewinder 21, drop is complete. We're pulling off to refuel."

"Copy Sidewinder, thanks for the heads up. See you shortly," I responded, acknowledging their successful drop and upcoming refueling. The seamless communication between the ground and air crews underscored the efficiency of our operations. As Sidewinder 21 headed for refueling, I began preparing for their return, ready to manage the next phase of our operation: Nighttime.

Night airborne operations were also slated in our training agenda, carrying the daytime drills into the realm of darkness with the added complexity of limited visibility. Although the use of night vision goggles presents a significant advantage, they are not without drawbacks, notably the phenomenon of tunnel vision that can restrict peripheral awareness. In these operations, aircraft navigate without the use of any external lights, mimicking the conditions they would face in actual combat scenarios. This approach is encapsulated in the maxim "we train as we fight," which stresses the importance of realistic training conditions to prepare for the rigors and uncertainties of combat with as much fidelity as possible.

Both helicopters took to the air just after dark. They notified us of their approach and readiness to resume. I called Staff Sergeant English to let him know.

"Hey Staff Sergeant, Strike and Sidewinder on the radio. Both are in bound."

"Tracking," he responded, "take the lead and guide them in. The LZ is ready. If there's any issues, I'll be around, but I'm sure you've got it."

"Roger that, Staff Sergeant. On it," I confirmed.

Little did Strike 23 know that the voice on the radio managing the drop zone belonged to a twenty-year-old boy, fresh out of high school, embarking on his first venture away from home.

With the task now mine, thanks to hours of observing and heeding Staff Sergeant English's guidance, a sense of profound realization washed over me. How did I, a mere youngster from a small Iowa town, end up with such significant responsibility in my grasp? Yet in that moment, I felt an unprecedented surge of strength and confidence. For the first time, everything clicked; I was truly in flow.

Staff Sergeant English had placed his trust in me, empowering me. I comprehended the enormity of the responsibility. Determined not to let him down, I embraced the role with all the vigor and dedication I could muster.

The Benefits of Delegation

By delegating tasks and responsibilities, you can help your team members develop their leadership skills and gain valuable experience. Unfortunately, many lower-level employees are not given these opportunities to grow, and, as a result, struggle when they are promoted to leadership positions. By providing these experiences earlier on, you can equip your team members for success before they even step into a leadership role.

Those in positions of power often have a strong desire to maintain control over the process, often driven by their ego. They believe that they are the only ones capable of doing things correctly and prefer to have things done their way. However, you must recognize that you may not be the most knowledgeable person on your team. Your subordinates could come up with better and more efficient solutions. By delegating, you open yourself up to new ideas and alternative approaches that you may not have considered before. While it may be challenging to let go of control, the potential benefits that your team can bring to the table make it worthwhile.

Delegating also cultivates a sense of ownership within an organization. When your team members are involved in the decision-making process, they are more likely to be invested in the outcomes. This sense of ownership grants the entire team a voice in shaping how things are done in the company, reducing complaints about the process since they are the ones who helped create it.

It will also lighten your workload. I often told my team, "If you are feeling overwhelmed, you probably aren't delegating enough." When you delegate, you are demonstrating that you have faith in your team members' capabilities and their potential to perform well. This trust is often reciprocated with a sense of loyalty, which can contribute to a positive and collaborative work environment.

What Does it Mean to Mentor?

Mentoring is the process of building new leaders in your organization. You are preparing them to be successful in future positions by guiding, motivating, and educating them. Every single person in your organization needs mentoring. You are also mentoring subordinates on how to mentor their subordinates through example and education. Don't wait until a position needs to be filled before you start to train the replacement. The military always has people coming and going. The big Army is moving people around constantly. This can be beneficial in the overall picture as it spreads knowledge and experience. In any dynamic environment, it's crucial for everyone to be ready to step into a leadership role at a moment's notice, particularly during times of chaos. You're not just expected to fill the shoes of a leader, you must excel in them. This necessitates an ongoing process of mentoring and preparing potential leaders in real-time. To ensure readiness, we regularly engage in drills that simulate scenarios such as the sudden loss of a leader in action. The next person in line must rise to the occasion, and so must every individual down the chain of command. Reflect on this: what kinds of similar training exercises could you implement in your organization to foster this level of preparedness and adaptability?

The 'One Up Drill'

"Staff Sergeant Waters is down, you're leading now. What's your next move?" This simulation was a staple in our training regimen. It posed a critical question: *If your superior was suddenly gone, how would you respond?* This exercise thrust participants into a leadership role, challenging them to take the initiative. It provided a valuable opportunity for learning and showcasing leadership capabilities. One-Up drills served as a powerful method for cultivating future leaders, equipping them with the skills necessary for combat readiness. It also gave them perspective on the stresses of leadership.

How to Properly Delegate

How do we effectively delegate? Begin by going through your daily list of tasks and asking this question: *What can I pass on that others could learn from?* Avoid delegating tasks simply to offload work that you don't want to do yourself. Delegating in this way can create resentment and damage your team's morale. Instead, focus on identifying tasks that could benefit from the perspectives and skills of others and provide valuable learning opportunities. Be honest with yourself when determining which tasks to delegate and which ones to keep. You can create a positive and productive work environment where everyone has the chance to contribute and grow.

Passing on the authority to make decisions doesn't rid you of the responsibility to hit deadlines and maintain excellent work quality. If you pass authority and they mess up, it's on you. They still need the appropriate training and mentorship. That is your responsibility. It could be your intent to give them very little guidance to see what they can come up with, but the lack of guidance must be deliberate: A tool to mentor. Again, I want to emphasize the importance of giving subordinates the power to do the task. It's not fire and forget. You can't just throw the task at them and then leave them completely out to dry, not knowing what's going on. And it's your job to figure out what that's supposed to look like. It's your job to assess.

It's important to remember that every team member has their own unique strengths and areas for improvement. You must have a clear understanding of each person's needs in terms of mentoring and assistance. While some team members may require more guidance and support, others may be more self-sufficient and prefer a hands-off approach. By understanding these differences, you can tailor your leadership style to provide the right level of support for each team member. This can help you build stronger relationships with your team, foster trust, and respect, and ultimately help your team members grow and develop into more effective contributors. So, take the time to evaluate each team member's needs and preferences and adjust your leadership approach accordingly.

You don't want to hover. You want to provide just the right amount of mentoring. Imagine being delegated to and told to make it happen. Then every step of the way, there's your boss just looking over your shoulder, "Oh, that's not how you should do that. No, no, we don't want to do it that way. Here's how you need to do it." It would be frustrating. In fact, you wouldn't want them to delegate anything to you. Delegation directly connects with the Commander's Intent concept. When we give someone a task, we're empowering them, making sure the Commander's Intent has been clearly communicated. They know exactly what the outcome needs to be. Be sure you're deliberate in how much gray area you are giving them to figure out.

Hopefully, they will come to you with questions because there's an open line of communication. You will encourage them to try to figure things out independently, providing the direction but not the steps. Have them figure it out. Get them to be creative in their solutions. They don't need to do it the same way you did it or the same way somebody else did it.

Conversely, there are likely tasks or projects your boss handles that you believe you could perform and even improve. You may have ideas to enhance their efficiency, especially if they're part of your daily routine. With your intimate knowledge of the process and a comprehensive understanding of all the moving parts, taking ownership can effectively lighten your boss's workload. This is an excellent step in your professional growth and a key aspect of the mentor-mentee relationship.

Do's and Don'ts of Delegation

There are things that can go wrong in the process of delegation. However, keeping these do's and don'ts in mind can keep the road clear for you.

Do

- Assess what tasks or projects are appropriate for the skill level of the person executing them.

- Encourage your subordinates to delegate appropriate projects.

- Empower your team to make the decisions needed to complete the project.

- Have open lines of communication and be very clear on expectations.

Don't

- Delegate complete responsibility.

- Be overbearing on how the task is completed.

- Push your "shit work" onto someone else.

It is time for you to start thinking about how you can involve your subordinates. What tasks or projects can you assign them? As a leader, are you mentoring your entire team and preparing them as much as possible? Be deliberate!

KEY POINTS

- Get some weight off your plate—delegation isn't dumping your crap, it's building leaders.

- Build leaders every day—don't wait for a crisis to develop them.

- Know your team's strengths and needs—mentor them with that in mind.

- Push your team to delegate as well—empower others to step up.

- Pass authority but stay responsible—make sure they've got what they need to succeed.

- Use drills like "one-up" to get people thinking beyond their roles and prepare them for leadership.

6
Watch Your Tone

"People may hear your words, but they feel your attitude."
– John C. Maxwell

During my final visit back home before my first combat deployment, I had a conversation with my father that I will always remember. His parting words resonated deeply, "I wish I could stand right there beside you." In an instant, I imagined my father as a fellow soldier. What would his journey be like? How would he navigate the challenges of military life?

I couldn't escape the thought of how my father might be treated in the military. There were many occasions throughout my time in the military when I was treated not just harshly but shown disrespect for no apparent reason. I imagined my father enduring the incidents I'd experienced. It struck a chord. Though I acknowledge the necessity of stern leadership, being pushed beyond limits, and enduring hardships, I didn't like the idea of my father being disrespected.

My memories of military discipline came to mind—the demanding superiors, the rigorous trials, the instances of being yelled at. These elements I anticipated and even embraced. I knew they shaped me into a stronger individual. But the concept of being treated with disrespect didn't sit right. I believe that molding resilience and determination can be achieved without demeaning another person. After experiencing disrespect, it becomes challenging to muster the desire to follow the individual responsible for it.

From these experiences, I developed my own principles: how I envisioned treating my family member—whether father, mother, brother, son. It became clear how I wanted to lead: a commitment to treating all with respect. I resolved that I would never disrespect anyone regardless of the circumstances. Don't get me wrong, I would often be straightforward, assertive, and at times seemingly aggressive.

However, I would never lead with demeaning conduct and would always strive to maintain respect. While it may be challenging at times, especially when the individual receiving the feedback may not appreciate it, my intention is always to benefit the team or help align their behavior with their potential.

I realized that every person I encountered, no matter their rank or role, was someone's family member. Brothers, sisters, sons, daughters—each held a place in someone's heart, just as my father does in mine.

The nature of our profession demanded a certain fortitude, a toughening of the spirit. Yet, this sternness should not eclipse our humanity. The line between cultivating strength and nurturing compassion became clearer than ever. The challenges we posed to one another need not undermine the profound respect we hold for each other's dignity.

I would always do my utmost to be respectful, and understanding, and remember that we're all someone's cherished loved one. The way I treated every soldier I encountered set the tone for how I wanted my company to be run.

What Is Organizational Tone?

The organizational tone is the projected attitude towards any **subject, situation, person**, or **place**. I use the word "projected" because it is possible to have a less-than-positive attitude towards something, but your outward display remains positive. You can still set a positive tone. It shapes the atmosphere of a place, setting the mood. When discussing a topic, the speaker's feelings about it become evident, as they project their attitude toward the subject. The tone isn't always how a person speaks; in fact, it can even be how a person makes you feel when they enter a room. How they carry themselves, their gestures, and their eye contact.

This tone might take on characteristics such as:

Cheerful – Serious – Assertive – Optimistic
Professional – Informal – Frustrated – Anxious
Critical – Lighthearted – Threatening – Frantic

Insert any number of adjectives and they could be used to describe your organization.

Anyone can display a tone or attitude towards their **work**, **team**, or **organization**.

Tone Towards Work

What attitude do you carry about your work? Everyone notices the presence of a discontented boss.

Think of a high school teacher who seems miserable in their role. Every day seems like a burden and chore for them. Complaints about insufficient supplies, difficult students, and the challenges of unreasonable parents fill the air. Their demeanor appears weary and defeated. Yet, this is the very individual entrusted with inspiring others to push their boundaries and view life optimistically.

Now compare that to a teacher who thrives in their role. They embrace every day with enthusiasm and dedication. They creatively overcome obstacles, addressing them with innovative solutions. Their demeanor is upbeat and resilient, clearly reflected in their teaching. This is the person charged with motivating students to exceed expectations and approach life with a positive outlook. That is the person we want in charge.

Tone Towards Team

How you treat your team and speak to them reflects your tone. This encompasses your attitude towards your team members and all your interactions with them. Do you speak to members of the team like they are idiots or less important than you or do you treat them with respect, encourage open communication, and value their contributions?

Staff Sergeant Hister and Staff Sergeant Neal were both squad leaders in my platoon when I was a young private. Both held comparable expectations and standards for their squads, both demanding exceptional performance and an unwavering commitment to achieving superior outcomes. While they shared a competitive spirit, their approaches to interacting with their teams differed significantly.

Hister excelled as a mentor, displaying genuine concern for each member of his squad. He interacted with his team in a relatable manner, treating them as peers. In contrast, Neal consistently employed a condescending tone during interactions, often resorting to belittling humor. The banter was one-sided due to the rank structure. So, it was unadvisable for subordinates to reciprocate. This treatment resulted in the completely different tones of the two opposing squads.

Hister's leadership philosophy emphasized a two-way exchange of expectations. He not only held high demands for his team but also encouraged them to expect the same from him. He exhibited a profound comprehension of his team members' personal and professional motivations, addressing challenges both at work and at home. This approach fostered a strong bond within the team. Care for each other extended beyond work-related matters.

In contrast, Neal's leadership approach lacked authentic connection. While he shared humor with peers, he struggled to establish meaningful conversations with his subordinates. This absence of meaningful engagement resulted in a team that lacked enthusiasm and struggled to meet expected standards, which caused even more frustration.

Hister's team demonstrated their dedication due to his genuine concern, resulting in a strong work ethic and adherence to high standards. On the other hand, Neal's team lacked the same level of enthusiasm. While Neal possessed solid technical skills, his approach often mixed valuable insights with unnecessary ridicule.

Staff Sergeant's Hister and Neal shared expectations for excellence, but their leadership approaches diverged significantly. Hister's mentorship and genuine care built a devoted and high-performing team, while Neal's disconnected leadership style led to a lackluster and less motivated group.

Tone Towards Organization

There may be resentment within a company stemming from how team members perceive their treatment or the treatment of their colleagues. This resentment can arise from various factors, including working conditions, inadequate opportunities to voice concerns, insufficient time off, or subpar benefits. As a leader, your responsibility is to identify these issues and assess their validity. More importantly, it's crucial to maintain open and transparent communication about the steps that will be taken to address them. This approach may not always align with what employees want to hear, but honesty and upfront communication will garner respect. Avoid sugarcoating the situation because people are often more perceptive than we give them credit for. They possess a finely tuned "bullshit meter" and can discern when they are being misled, or the core issue is not being directly addressed.

We've all encountered situations where individuals talk in circles to evade addressing the actual question. This is particularly evident in politics. In the corporate context, avoiding this kind of obfuscation and engaging in honest, direct communication can help build trust and foster a more productive and harmonious work environment.

Tone Is Amplified Under Pressure

Your team closely observes your behavior under pressure, noting how you handle situations. Do you become anxious, snap at people, and make impulsive decisions? Your reactions during challenging moments define the tone you set in such situations.

You are an example of what the tone should be. You must bring it every day. Be the example of what you want to see. You want your team to care about your organization like it is their family, you better care about them like they are your family.

I had a reputation for maintaining an unusually calm demeanor even in stressful combat situations. Patrolling in our MATV (Medium Armored Tactical Vehicles) through the harsh terrain of Afghanistan, particularly the treacherous thirty-mile stretch from Asadabad to our battalion headquarters, there was always the risk of ambushes and attacks. This route had already claimed the lives of some of our comrades, and the weight of that knowledge never left us. It was a road that demanded seriousness.

One afternoon as we traversed this danger zone, we were suddenly ambushed by heavy machine gun fire and rocket-propelled grenades. The Taliban strategically fired from elevated positions in the foothills and concealed spots in the nearby cornfields. In such ambush scenarios, there were two general approaches to consider. Sometimes, you pressed forward, recognizing that the enemy had set the stage for this confrontation, and the odds might favor them. Other times, you would find the nearest cover, shielding yourself from the incoming fire, and return fire with every available asset. A key advantage was the unpredictability of your response. The enemy never knew what your response would be.

On that day, we opted to bring the full force. As our platoon retaliated, I took control of the radio, calling for additional support. I explored all possibilities, from mortar fire originating from the nearby friendly base to artillery from a more distant location. Then I considered the potential for Apache or Kiowa helicopters and, if the Air Force wasn't preoccupied elsewhere, the possibility of F16s or, my personal favorite, the A10 Warthog. Fortunately, we managed to secure mortars, artillery, and Apache helicopters.

One significant challenge in such situations was coordination. You couldn't have mortars or artillery shells in the air when helicopters were present. It required precise timing, knowing when to deploy each asset and when to shift focus. It was an orchestration of controlled chaos.

While we engaged in a fierce firefight our platoon leader constantly adjusted our positions to outmaneuver the enemy. I remained on the radio, switching between frequencies, communicating with different aircraft and mortar teams. I carefully directed mortar rounds and guided the Apache helicopters to their targets using Hellfire missiles. The battle persisted for nearly an hour, but eventually, we neutralized the enemy, and the firing ceased. We were fortunate to have escaped that day without serious casualties.

Upon our return to the Forward Operating Base, we dismounted from our vehicles, taking the opportunity to inspect the damage we had sustained. Tony Salinas, our platoon leader, approached me with an ear-to-ear grin. He exclaimed, "Holy shit that was close." He went on, "Hearing you over the radio is funny. All hell's breaking loose and your voice remains as calm as ever." Tony and I both understood the importance of this composure. If the guys heard panic in my voice over the radio, it would only add to their fear and uncertainty. By this time in my career, I had been in too many firefights to count. But no matter how many times you've faced enemy fire, the fear never truly dissipates, especially when the lives of your team depend on your decisions.

Maintaining composure wasn't just about reassuring the team that the situation was under control, it also ensured that the assets I was coordinating understood their instructions clearly and trusted that they wouldn't be sent into harm's way. Imagine the opposite: a voice filled with fear and indecision. The impact on everyone involved could have been catastrophic.

When chaos reigns, be the eye of the storm.

Steps to Controlling the Tone

1. Lead by Example.

First and most importantly, be the example you expect from your team. Demonstrate the behaviors, attitudes, and communication styles that you expect in others.

2. Get in the Trenches.

You will never truly know the tone of your team if you never leave your office. You need to establish a strong connection with your team at every possible level. It's essential to engage with them and understand their challenges and difficulties. Your goal should be to create an environment where they feel comfortable enough to express themselves honestly. They will engage with you if they perceive you as a leader who actively resolves issues. So, get your ass in the trenches and know it's more than just showing up; it involves engaging in meaningful two-way dialog with individual people. Actively seek their input and solutions to the challenges they encounter on a daily basis. You're not merely a visible presence; you're dedicated to fostering genuine relationships. Superficial gestures like a thoughtless pat on the back won't enhance your influence. You must genuinely care about their concerns. Inauthentic interactions can even harm relationships. You don't have to be friends with everyone on your team, but they do need to respect you.

3. Identify Influential Figures.

Identify the individuals who hold sway over the tone of the team. Those who most regularly interact with the people in the trenches. These influencers play a significant role in shaping the atmosphere. If they are exemplary contributors, you must provide them with encouragement and support. However, if they exhibit a persistently negative attitude, address the issue promptly. Treat the negative attitude as if you're uprooting a problematic seed that could take hold and spread throughout your organization. It is your responsibility to address issues head-on.

Be frank and make sure everyone is aware of your expectations. They must know that you expect them to perform at high levels. Give genuine and transparent feedback when necessary. Encourage others to step up and become positive influencers. Identify potential leaders at various levels within the organization and support their efforts to promote a better tone.

4. Be Consistent.

Ensure consistency in tone-setting efforts throughout the organization. This means holding everyone, including leadership and influential figures, accountable for maintaining the desired tone. If you say one thing and do another, you are creating an appearance of a double standard. Continuously solicit feedback from your team.

Let leaders know it's okay to voice constructive criticisms to a superior as long as subordinates are not present. All of the organization's leadership needs to appear united, not divided.

5. Always Adapt.

Be open to adjustments in your approach. The process of changing the organizational tone is constant and ongoing. It's a gradual process that requires commitment, patience, and consistent efforts from leaders and influential figures at all levels.

Climate Surveys

Do Organizational Climate Surveys Work? Absolutely, yes. In fact, I strongly encourage you to utilize them, especially if your organization is large or you have a widespread presence that makes it challenging to be involved in every aspect. Climate surveys can serve as an effective tool but remember that they are not all-encompassing.

The crucial aspect of climate surveys is that they will only be beneficial in boosting morale if you not only read them but also act on the issues raised. If you choose not to act on the feedback, they could potentially do more harm than good. If you're only pretending to care about the concerns your team has, it is blatantly obvious.

In conclusion, it's crucial to recognize that tone permeates every aspect of human interaction. It molds perceptions and shapes attitudes. Whether you're dealing with work-related matters, interacting with team members, or navigating high-pressure situations, the tone you project speaks volumes and molds the world around you.

Always keep in mind that your demeanor sets the precedent. If you convey frustration and negativity, those around you will mirror that sentiment. Conversely, if you remain composed under pressure, it encourages others to do the same. When you communicate with clarity and directness, you inspire others to follow suit.

Engaging with your team in the trenches offers a perspective that influences your effectiveness as a leader. Treating your team as if they are your clients and embracing a leadership attitude of service is guaranteed to build a strong supportive team. This approach initiates a ripple effect throughout the organization.

Remember, when it comes to tone, you will get what you tolerate.

KEY POINTS

- Lead by example—your behavior sets the tone for the whole team.

- Speak positively about your job and your team—don't allow a culture of negativity to take root.

- Respect everyone like they're family—cut the demeaning crap.

- Keep it real—speak honestly and upfront; talk to others like you want to be talked to.

- Remember, tone is amplified under pressure—practice responding effectively in high-pressure situations to maintain control.

7

Rewards and Punishments

"Rewarding hard work is one of the most effective ways
to reinforce effort and commitment."
– John C. Maxwell

My military career and the raw realities of combat taught me more about human nature and motivation than any book or classroom ever could. I saw soldiers push beyond their limits. I saw the potential for broken spirits. I witnessed the pride of many men and women when they received well-deserved medals, and I saw others receive equally deserved reprimands.

Although the military's system of rewards and punishments may seem stringent and harsh, it is effective. More than just maintaining discipline, the goal is to ensure each unit and each soldier can function cohesively under extreme stress when the stakes are unimaginably high.

The principles we live by in the military are not confined to the battlefield. Nearly every civilian organization faces the challenge of leading and motivating their people. How do you inspire a team to rally behind a vision? How do you cultivate a sense of responsibility and drive? The military's reward and punishment approach provides valuable insight into reinforcing positive behavior and motivating teams. Motivation requires a foundation of genuine appreciation and accountability. It is remarkable how often organizations overlook these elements and create lackluster environments where team members go unappreciated and individuals evade responsibility.

I want to share years of wisdom passed down to me from experienced military leaders on the impact and significance of rewards and punishments. Keep in mind that the heart of leadership, whether on the battlefield or in the boardroom, lies in understanding the people you lead.

Why do rewards matter in the grand scheme of leadership and organizational success?

Five reasons why recognizing and rewarding is important.

1. Motivation and Productivity

2. Team Engagement

3. Retention

4. Improved Performance

5. Team Morale and Collaboration

The Science of Reinforcement Theory

The relentless pursuit of excellence is at the core of military training. Reinforcement theory demonstrates that when individuals receive the recognition they've earned, it is a shot of adrenaline. This reinforces the behavior and makes people want to keep pushing their limits.

B.F. Skinner, an American psychologist and behaviorist, is regarded as the father of Operant Conditioning—a method of instruction that links behaviors with consequences. He was a professor at Harvard from 1958 to 1974 and has written several books on the subject. He even worked on a secret project to train pigeons to guide bombs during WWII.

Skinner's core belief was that positively reinforced behavior tends to recur, whereas behavior met with negative outcomes is less likely to be repeated. In essence, if you are rewarded for a behavior (good or bad), you will continue to do it. Skinner outlined three possible responses you can have to a behavior. You can reinforce behavior, punish behavior, or remain neutral toward the behavior.

1. The Reinforcer Response

Any response that increases the probability of a behavior being repeated. Reinforcement can be for positive or negative behavior.

A reinforcing response to positive behavior might be as simple as an instructor commending a student for proactively helping a peer understand a tough concept. They're not just one-off rewards but tools to encourage lasting positive habits.

To understand a reinforcing response to a negative behavior let's think about a situation in which a sergeant frequently bypasses the chain of command by going directly to the company commander with concerns, rather than addressing them with his immediate superior, a lieutenant. Instead of reprimanding the sergeant for this breach of protocol, the company commander entertained his concerns and even praised him for his "directness" and "initiative". This recognition made the sergeant feel empowered and validated, leading him to continue sidestepping the lieutenant. Over time, this erodes the integrity of the chain of command and undermines the lieutenant's authority, all because the initial behavior was rewarded rather than corrected.

2. The Punisher Response

A response that decreases the probability of a behavior being repeated. The behavior may be positive or negative.

How do you use a punisher response for positive behavior? If, in the above example, the sergeant was reprimanded for disregarding his chain of command, he would be less likely to repeat it. Resulting in a positive learned outcome by punishing the negative behavior. On the other hand, sometimes positive behavior is met with a negative response. If a sergeant spoke up in a mission planning session with an idea he thought would save time and he was met with a response of "That's not how we do it here," this response would stifle team input.

3. The Neutral Operant

A neutral response neither increases nor decreases the probability of a behavior being repeated.

Skinner argued that if you don't reward behavior, it's likely to vanish. This is where I believe many leaders can improve. It's crucial that dedication is acknowledged and celebrated. Imagine a team pouring their heart and soul into achieving targets, but never getting a nod of appreciation. I had a commander who would always emphasize that if behavior isn't rewarded, it might just fade away. I've seen the truth of this firsthand in the trenches. My platoon would tirelessly patrol, getting into firefights with the enemy, and risking our lives to complete missions. But there were times when our efforts went seemingly unnoticed. The fire in the soldier's eyes started dimming. How long can soldiers push themselves without a simple "good job"? They needed to know they were making a difference and not risking their lives for nothing.

Is consistently rewarding the same as bribery? I recall overhearing one of my soldiers jokingly say, "Medals are just bribery for bravery." It's a thought that has crossed my mind too. But is it inherently bad to expect recognition?

While it can seem that way, expecting recognition is not always negative. The real challenge is when that shiny medal becomes more enticing than the reason we're fighting. In the military, it's vital to ensure that every recognition is a testament to our larger duty and to anchor every reward to the bigger mission. And to avoid having so many medals that their value diminishes.

There are many options available when it comes to acknowledging and appreciating your team's efforts. This could be individual or team recognition. Regardless, be sure **you put as much thought and effort into the recognition as your team did to receive it.** What does it imply when they're working hard for the team, and you casually give them a ChapStick and a "you're the balm" note as a token of appreciation? You expect a high effort in their work, they should get a higher effort in the appreciation shown. Lack of effort implies a double standard, and it's truly disrespectful.

Individual Rewards

It's crucial to recognize that individuals within your team have unique motivations, and their reactions to rewards may vary significantly. Acknowledge each person's contributions in a way that's meaningful to them and builds up a sense of accomplishment and belonging. Find the right fit for each person, making sure they feel appreciated.

Some may find monetary bonuses or promotions particularly motivating, while others may be more appreciative of non-monetary recognition, such as public praise, flexible work hours, or opportunities for professional development.

By incorporating a wide range of reward mechanisms, you can create a more inclusive and effective recognition system. This approach recognizes that each team member's contributions should be acknowledged in ways that resonate with their personality and aspirations.

Reward the Team

Another approach would be to recognize the whole team for their outstanding work. Rewards not only boost individual morale but also strengthen team cohesion. The military often uses team rewards during friendly group competitions. If it's possible to turn it into a competition, we do it. Football becomes "Combat Football," we compete in ultimate frisbee, run ten-mile relay races, and complete the more obvious weapons proficiency competitions. Contrary to popular belief, the competitive disassembling and reassembling of weapons isn't just a movie scene, we frequently participate in such challenges.

"Mike," I whispered to my buddy beside me, "how confident are you with the 249?"

He shot me a sideways glance, the kind that screamed, "Are you kidding me right now?"

The sergeant's voice boomed. "Alright, boys! Today we're gonna see which team has been paying attention. It's a test of speed and knowledge."

I glanced ahead at the rubber container before us. It was filled with weapon parts, a mishmash of metal and precision, all jumbled up, waiting to be put together.

The whistle pierced the air, signaling the start. My mind focused. I sprinted twenty yards, my boots pounding the ground, and reached the container, plunging my hands into the icy metal of weapon parts.

The M249 Squad Automatic Weapon. I recognized its parts instantly, having spent countless hours disassembling and reassembling it. With precision and speed, I began piecing it together. Every second counted. The clinking sounds of metal echoed around me as other teams scrambled to assemble their weapons.

Just as I was snapping the last piece into place, I heard a shout from a neighboring team. "Wrong part, Danny! That's from the 240!"

I stifled a grin.

With the M249 perfectly assembled, I sprinted back, tagging Mike, who shot off like a bullet towards the container. I could hear the cheers of our teammates, the chants urging us on.

Mike was back in record time, having masterfully assembled the next weapon. We were in the lead, but barely. The team next *to us was hot on our heels, their third member sprinting towards the container even as Mike set our second weapon down.*

It went on, a blur of metal, sweat, and sheer determination. The air was thick with anticipation. And then, as our last weapon clicked into place and our final teammate sprinted across the finish line, the whistle blew again.

Victory!

We erupted in cheers, even hoisting our last member into the air. We made an absolute spectacle out of our celebration. For such a small event that took less than twenty minutes, we treated it like we had just won the World Cup.

It wasn't just about being the fastest or the most knowledgeable. It was about teamwork, trust, and knowing that when the pressure was on, we had each other's backs. It gave a hint of how each would perform when the stress was real: when one of these weapons would jam in combat and the lives of others were on the line. This modest contest seemed trivial, but it held immense importance to us.

Awarding Tactfully

Consider the wording when presenting team awards. What is the award truly acknowledging?

- Awards shouldn't be given haphazardly. It should only be when you spot the team exceeding expectations.

- It's essential to recognize all team members, whether they rank first or twentieth in performance. But when giving an award, it's also crucial to highlight members who are giving extra effort. For instance, "Many in the team work hard every day, and today I'd like to highlight Private Hardy for his outstanding effort."

- Ensure a variety of team members receive recognition and avoid favoritism.

- Have official events that are a great opportunity to show appreciation for your team. There are also great opportunities for you to single out individuals in front of the group for the work they are doing.

Here is a list of ideas for acknowledging and rewarding your team.

Custom Awards: Tailor-made trophies or plaques featuring the employee's name and their accomplishment.

Immediate Cash Bonuses: Unexpected financial rewards for exceptional performance or efforts.

Day off Pass: Grant an additional day off with pay.

Continuing Education Sponsorship: Cover the cost of a selected course or seminar.

Gift Cards: Gift cards for beloved eateries, cafés, or online retailers.

Spotlight in Company Newsletter: Showcase their success in the organization-wide newsletter.

Getaway Rewards: Offers for a leisure weekend or brief holiday.

Office Space Upgrade: Improvements to their work area, like a premium chair, desk, or tech tool.

Monthly MVP Recognition: Monthly accolades for the standout employee.

Dinner with the Boss: An informal meal with the company's executives.

Prime Parking: Exclusive parking space for the monthly top performer.

Team Acknowledgment: Public praise or acknowledgment during a staff gathering.

Group Honor Event: A celebration meal or outing to honor their achievements as a team.

Flexible Dress Code Coupon: Permission to dress down for a day of their choosing.

Fitness Club Memberships: Annual membership at a local fitness center.

Service Milestone Tokens: Gifts acknowledging service duration with the company.

Wall of Fame: A dedicated office area to exhibit hard workers.

Well-being Incentives: Offers for relaxation and health, like spa visits or massages.

Branded Merchandise: Personalized company gear such as apparel, drinkware, or accessories.

Movies Passes: Tickets to a current hit movie.

Special Projects: The chance to lead or partake in pivotal projects.

Online Courses: Membership to e-learning sites like Coursera, Udemy, or Masterclass.

Team Building Experiences: Team activities such as escape rooms or adventure sports.

Birthday Festivities: Arranging an unexpected party or cake for their birthday.

Inspirational Reads: A top-rated book in their specialty or a motivational work.

Discipline

In the regimented world of the military, discipline is paramount. From the first day at boot camp, when the drill sergeants' shouts echo in the ears of new recruits, to the demanding missions in hostile territories, we are constantly reminded of the consequences of our actions. For many in uniform, this strict code is not just about following orders, it's a matter of life and death. Every soldier knows the weight of responsibility and the potential repercussions of a lapse in judgment.

Immediate consequences have their place. When a soldier fumbles with their weapon during a rehearsal, the immediate reprimand ensures they won't make the same mistake in the heat of battle. In high-stakes situations, swift corrective action seems justified and effective. However, because of the complexities of human emotion and psyche, not every behavioral issue can be addressed with a mere reprimand.

The deeper challenges often require understanding and introspection, not just discipline.

Compassion is the antithesis of punishment. We must approach others with compassion, understanding, and genuine interest. Punishments tend to create divides; they label one party as the wrongdoer and the other as the enforcer. Compassionate correction, on the other hand, seeks unity, understanding, and collective growth.

Before jumping to conclusions about someone's performance, it's essential to consider various factors that might be influencing their work.

Here are the three things to ask yourself before administering any type of punishment:

1. Were they clear about the assigned task?

2. Did they receive the necessary training to execute the task?

3. Could personal circumstances be impacting their ability to work effectively?

Imagine the regret you'd feel if you sternly reprimanded someone for missing a deadline, only to later learn they were dealing with their spouse's cancer diagnosis or were stressed about overdue mortgage payments. Remember, there's often more to the story than meets the eye. Instead of making hasty judgments, approach such situations with empathy and understanding.

I was standing outside the company one day when a young man, around his late twenties, approached me. He had a freshly trimmed haircut and an evident nervousness in his eyes. Though he didn't seem completely unfamiliar with the Army, there was a hint of apprehension in his demeanor. As he came closer, he inquired, "Sergeant, can you tell me where the First Sergeant's office is?"

"Certainly," I replied, "What's your name?"

"Hernandez," he responded.

"I'll walk you there."

During our brief two-hundred-yard stroll across the compound, Hernandez shared that he had been AWOL for the past five years and had returned to "Set things right."

"Five years? And you've chosen to come back now? You're aware we're deploying to Iraq in five months?" I said, somewhat surprised.

"Yes, I've heard," he acknowledged.

Upon reaching the office, I knocked on the First Sergeant's door and informed him that Private Hernandez wished to speak with him. Leaving Hernandez, who seemed more anxious than before, I exited as the First Sergeant beckoned him in.

I knew how challenging the situation must be for him. Going AWOL, especially for a duration as long as five years, is a rare and serious offense in the military. The consequences can be severe, with soldiers facing potential imprisonment for such an act. Later that day, I mentioned the situation to a group of my peers, knowing there was potential he would be assigned to our team. The reactions were strong and varied.

One of them immediately said, "Send him to jail. If he's been AWOL for that long, how can we trust him on our team during deployment? We can't risk having someone unreliable with us."

Another, looking for a middle ground, suggested, "Let him deploy with us but assign him all the shit tasks. This way, he faces the consequences of his actions, and the rest of the platoon doesn't have to deal with those tasks."

Yet another was adamant, "Just kick him out. We don't need this kind of distraction. There are more pressing matters to concentrate on, and we can't afford any disturbances right now."

The variety of opinions highlighted the complexity of the situation and the challenge the command faced in making a decision about Hernandez's future. I remember being torn myself, wondering if he should be given another chance or if we should strictly adhere to the rules.

The following day, we learned the backstory. Five years ago, Hernandez had returned to Puerto Rico due to his mother's critical illness. Tragically, she passed away shortly after his arrival. Although his leave was temporary, the sudden death of his mother left a significant void and responsibility in the family. Feeling the weight of these responsibilities, Hernandez believed he had to stay in Puerto Rico to support his family, leading to his prolonged absence. Hearing this, my heart ached for him. I realized that sometimes life throws us curve balls that are hard to navigate. And now, after half a decade, he was back.

Fast-forward five years. Private Hernandez, feeling a strong sense of duty, chose to face any repercussions and fulfill his commitment to the Army. The higher-ups decided to retain him, and he was assigned to our platoon for deployment. I was curious about how he would fit into our team, given all the whispers and judgments surrounding him.

Private Hernandez quickly established himself as one of the most dependable and competent gunners I've ever had the privilege of working with. His reliability was unparalleled; he executed every order without hesitation. Initially, many of us were quick to form opinions about Hernandez without truly understanding his past or the challenges he faced in his personal life. I admit, I was one of them. Had the command taken our initial recommendations to heart, we would have lost someone who proved to be an invaluable member of our team.

It's an important reminder to all of us: how often do we judge others without truly understanding their circumstances or the battles they're fighting internally? I've learned to take a step back and remember Hernandez's story before passing judgment on others.

Mass Punishment

Group accountability, often referred to as "mass punishment", though not my favorite term, can be an effective technique when executed correctly. There's value in peers holding each other accountable, whether it's soldiers in the military looking out for one another or employees policing their fellow employees. It's not a one-size-fits-all approach since the military and the civilian world have different operational styles, but the fundamental concept remains the same. If the failure of one person or even a portion of the team is having an adverse outcome on the entire group, then it stands to reason the team should be expected to police itself to some degree.

In the military, if one individual fails to meet their duties and expectations, it can have a significant impact on the entire unit's success. While top-down accountability is crucial, there's also a role for peer-driven accountability. For this to work, the organization must foster a culture where everyone is genuinely invested in its success. If people don't care about the organization's well-being, they won't care about an individual's performance.

If you use leadership principles effectively and build trust to create organizational buy-in, individuals will naturally feel a sense of responsibility for one another. They'll apply peer pressure to ensure that everyone fulfills their responsibilities. When someone falls short, you need to take corrective actions.

First and foremost, transparency is key. As a leader, you must clearly communicate why everyone is being held accountable for one person's performance. Explain that the success of the organization depends on each member doing their part, and one person's failure can drag everyone down. In the military, this could have life-or-death consequences. Make the stakes clear.

When people understand this, they're more likely to hold each other accountable. If necessary, they may collectively decide to work extra hours to correct the situation. Not everyone will be pleased with the approach. However, leaders should prioritize the organization's success over being universally liked.

A bit of grumbling and complaining is normal, but it's essential to distinguish between those who vent their frustrations and those who consistently resist accountability. Individuals who continuously complain without recognizing the necessity for accountability might not be the best fit for the organization.

Accountability is especially important when a leader is not present. In a military context, when a soldier observes a peer engaging in unethical behavior, they should feel compelled to intervene and ensure the right thing is done. This kind of peer-driven accountability is crucial for the organization's overall success, as leaders can't be everywhere at once.

When used improperly, group punishment can lead to significant discord within the team and between the team and leadership. It's crucial not to hold the entire team accountable for the actions of a single individual if the team isn't responsible for the results of those actions.

The principles of rewards and punishments aren't just for the battlefield; they're critical in any organization. Recognition isn't just a nice-to-have; it's the fuel that drives dedication. And without consequences, standards crumble. Mastering these dynamics is a key to mastering your craft. Lead with genuine appreciation and accountability, and you'll forge a motivated, cohesive, and high-performing team.

KEY POINTS

- Reward good behavior every day—show genuine appreciation for the effort.

- Put real thought into recognition—make it meaningful and tailored to what drives them.

- Don't reward bad behavior or hand out awards like candy—make recognition count.

- Be fair and impartial when it's time to punish—don't play favorites.

- Be thoughtful with corrective actions—consider their personal struggles when handling mistakes.

8
Find Structure

"The function of organization is to eliminate confusion and disorder."

– UNKNOWN

TWO OF THE MOST VALUABLE THINGS to a human being are time and peace of mind. Being organized can give you and your team both. Leaders have the power to promote and foster the culture of their organization—if you are organized and expedient in your execution of tasks, then your team will learn these habits as well. Making responsibilities and the flow of work clear promotes a sense of purpose and mastery for everyone.

Conversely, a lack of organization can destroy your team's confidence in you. Without a clear structure or framework guiding your work (and life), you risk falling into a cycle of missed deadlines, reinventing processes, and repetitive errors.

Reflect on how you perceive disorganized individuals. Consider the impression you'd get entering an accountant's office cluttered with scattered papers and files. Would you trust them to meticulously handle your finances? Probably not. In the same way, if you don't implement an organized system as a leader, your team's efficiency will deteriorate. Forgetting important details or failing to provide clear guidance can lead your team to doubt your leadership capabilities. Effective leadership includes maintaining a structured and functional team.

People often envision the military in terms of high-octane moments portrayed in movies: The pulse-pounding rush of close-quarters combat or using shape charges to breach doors. Yet this isn't the everyday reality. Beneath the surface of these thrilling scenes lies a world of essential administrative functions—the unseen but vital heartbeat of our operations. For every door explosively breached in a training exercise, there are countless hours of planning, organizing, and preparing. It's in these unglamorous but equally crucial moments that an organization is shaped. While it's natural to gravitate towards the more interesting aspects of your job, do not neglect the mundane.

In this fast-paced competitive world, time is the ultimate resource, and harnessing it is a superpower. We always feel like we need more time to get things accomplished. Our enemy isn't getting any weaker. Our strongest competition isn't taking a rest day. Wasting time due to a lack of procedures is the last thing you want. You absolutely need to focus on the training and tasks that will increase the efficiency and effectiveness of your team. Stop wasting time and destroying the confidence of your team.

This chapter delves into creating a structured framework for your team. Being an organized leader is a big topic, one that is frequently undervalued and deserves more attention. A topic that could fill its own book.

Disorganization Leads to Chaos

A major pitfall of an unorganized team is the absence of defined processes. Without clear guidance on how to get things done and how the organization is structured, you have a ship without a rudder.

Job descriptions clarify each member's role and responsibilities. Organizational charts tell who to approach for specific questions. When you lack specificity in these areas, people just aimlessly wander. Soldiers without orders. You might be asking two different people for the same thing or find that two people are needlessly doing the same job. Even worse, you might ask five different people how to complete a single task and receive five different answers. When there is no consistency, your organization can be a complete shit show.

Disorganized teams suffer from a lack of confidence in their leaders and fellow team members. Moreover, it's unlikely that such a team can achieve its goals effectively. The remedy for this chaotic situation starts with introducing structure.

How To Create Structure

Begin by considering every task you and your team encounter daily. These include tasks that are done by you, your subordinates, or your boss. Now, categorize these tasks into two types: **technical** or **tactical**. Technical refers to tasks that produce a known outcome. They might be difficult or easy, but they have a standard outcome. Technical tasks may be repeated and scheduled because they are done the same way every time. Examples of technical tasks include ordering supplies, paying bills, or completing reports. Most administrative-type tasks are technical. Tactical tasks are non-standardized and can have various, sometimes unexpected outcomes. Tactical processes often require human reactions and responses—adaptability in the moment. Making sales calls, de-escalating angry customers, or deciding when to open fire on the enemy are all tactical tasks. While categorizing your tasks, everything you do will fall into one category or the other.

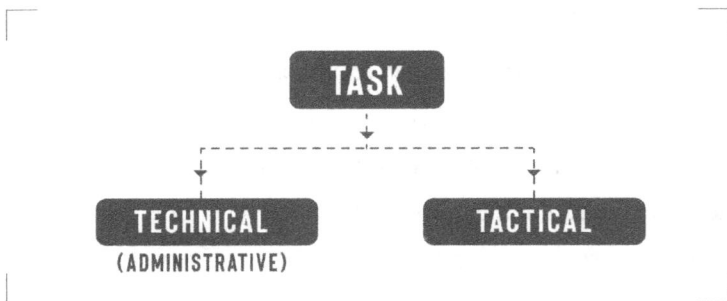

Figure 8-1

Let me give you an example of a technical task.

I taught the technical tasks associated with firearm ballistics to snipers and sniper trainees for many years. Any relatively intelligent person can learn how to do all the calculations required to shoot long distances. Obviously, some will be better than others, but anyone can learn it. There are numerous tasks and calculations required to do this. You must look through the scope of your rifle and use the dots in your scope in a quick equation to determine the distance to your target. You need to assess the wind—its speed, direction, and how it might impact the trajectory of your bullet. You may need to compare the altitude of your target to the altitude you zeroed your rifle. The thinner air has less resistance, therefore, it will affect the velocity of your round. You also need to determine the angle from which you are shooting. If you are on top of a mountain and the angle is steep, you must know how to figure the cosine to determine the real distance to your target.

With all these steps, I could teach you exactly how to get the perfect shot. With all the right calculations, each sniper should achieve relatively the same outcome.

Let me give another example we can all relate to. How about your taxes? In theory, you should be able to teach or be taught how to do taxes, and anybody could plug numbers into the formula and have the same outcome. This is what the online tax services are. You do not have to become a tax expert; you're plugging in numbers and you're coming up with an outcome. It is complicated but doable.

Conversely, we also deal with tactical tasks. These are scenarios where each action may lead to an unpredictable outcome. On the battlefield, for instance, you can't precisely foresee how the enemy will react to a particular maneuver. There are times when they will retreat and other times when they will advance. Their response can vary day by day.

Let's circle back to the sniper scenario. Sniping isn't just about precision shooting, it involves a hefty dose of tactics. Snipers need to pore over intelligence reports, using this information to predict enemy movements. This intel is crucial for deciding their insertion strategy. But it doesn't stop there. They must craft a plan, detailing every step from insertion, to movement, to taking the shot. This includes alternate routes and, if needed, setting up decoy positions.

Using the terrain to our advantage is key, but so is knowing the aftermath of the shot. What will the enemy's response be? If a sniper's technical skills (the meticulous calculation of the shot) are not spot on, they're left facing a pissed-off enemy that going to retaliate.

Despite extensive tactical training and countless rehearsed scenarios, predicting an enemy's reaction is never certain. Sure, superior training improves the odds of accurate anticipation and response, but remember, in the realm of tactics, nothing is guaranteed.

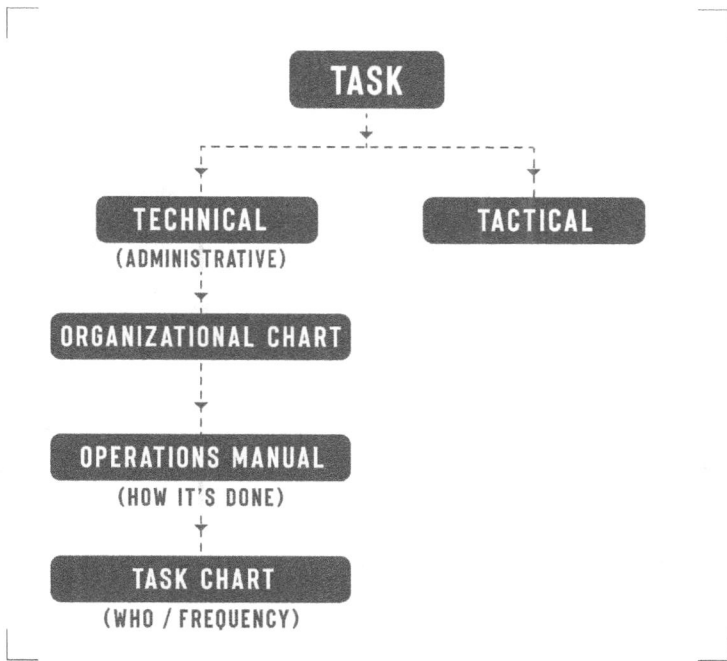

```
                          ┌──────────────┐
                          │     TASK     │
                          └──────────────┘
                 ┌────────────────┴────────────────┐
                 ▼                                  ▼
        ┌──────────────────┐              ┌──────────────────┐
        │    TECHNICAL     │              │    TACTICAL      │
        │  (ADMINISTRATIVE)│              └──────────────────┘
        └──────────────────┘
                 ▼
     ┌────────────────────────┐
     │  ORGANIZATIONAL CHART  │
     └────────────────────────┘
                 ▼
     ┌────────────────────────┐
     │   OPERATIONS MANUAL    │
     │     (HOW IT'S DONE)    │
     └────────────────────────┘
                 ▼
     ┌────────────────────────┐
     │      TASK CHART        │
     │   (WHO / FREQUENCY)    │
     └────────────────────────┘
```

Figure 8-2

Now that we've categorized your tasks into **technical** or **tactical**, the next step is to manage them. Starting with the technical tasks, the key to tackling these is to use a repeatable and systematic approach—straightforward and no-nonsense. Just as there's a meticulous, step-by-step process to execute a sniper shot, the same principle applies to your routine tasks.

Your goal is to establish a clear procedure for each technical task on your list. In the military, we often say, "Do routine things routinely." Develop an organized, step-by-step method for each task, making execution second nature.

Get Organized

You will first need a clear organizational chart. A simple chart that lets everyone know the line of authority and who is responsible for what. It will eliminate any confusion about job roles. There should be no guessing about responsibility.

Remember not to stifle cooperation and collaboration between your team and other teams. We want people to help each other out because some tasks will depend on more than one person. In such collaborative tasks, the level of communication and interaction is directly proportional to the quality of the outcomes. First, we must stick to our charting step and clearly define everyone's roles.

In my career, I learned the importance of the Keep It Simple Stupid (KISS) approach, and you should do the same. (Or at least keep it as simple as possible.) I also learned to balance simplicity with remembering not to be overly rigid in processes and projects. Organizations with little room for flexibility are not desirable workspaces. Flexibility makes room for growth and creativity. Therefore, I will not oversell the importance of systematic procedures and routines so much that you create a rigid and uncreative work atmosphere. However, you must find a productive balance between standardized procedure and creative freedom that suits your organization and work type. Don't let creativity be an excuse to shun a systemized structure in your team and organization.

Many leaders hesitate to standardize processes, they are reluctant to define the job roles of their employees or the regularity of their tasks. For the lazy leader, there is an attraction to non-standardized processes. They do not want a codified organizational chart because it holds them accountable to standards they don't like. Such leaders never leave good impressions or legacies because the quality of their tasks will always be compromised due to of their slothfulness. This prohibits leaders from creating respect and building high-functioning teams.

Establish a Systemized Approach

Your next step is to approach your tasks systematically. Imagine you are going on vacation for a month, and another team member is going to fill your position while you are gone. You would need to write down specific instructions on how to do everything you do and guide the replacing team member. Most tasks on your list will be technical stuff with a standardized outcome. You will write out the entire process as an instruction manual for your job. You are essentially standardizing the technical aspect of your job. Then do this throughout your whole organization after creating the organizational chart. Once it's put together, you have an operation manual that tells how your organization operates on a technical level. It should be a document you could hand to any new employee to read, and they would understand the technical workings of your organization. It also reduces the time required for the onboarding process.

Next, you must determine how often and when tasks are repeated. A task chart can be a helpful tool for organizing and charting in detail. It covers who completes the task and the frequency it is done. This makes responsibilities for the tasks clear—generally for routine tasks. A task for a one-time project would require its own task chart. Project management professionals often use a software called Gantt for task charts. There are many online examples to reference. Routine tasks done routinely will help to prevent anything from being overlooked and falling through the cracks. Again, you're writing this manual as if it were for your replacement.

Organizing tasks does more than putting things in order. It makes each task meaningful and sets priorities, turning every small task into a step toward a bigger goal. But the true value comes from taking a closer look. When we regularly review what we're doing, we think more about whether each task is really needed, this review process allows us to cut out what isn't necessary. Too often we do things just because that's how they've always been done.

Efficiency reduces the likelihood of dissatisfaction that employees face today: lack of meaningful work. Additionally, you would ensure that your team members are content and give their best performance. From the leader's perspective, task charts can be an invaluable tool for auditing and assessing the team's performance.

When assessing a civilian business, the physical business is of little value to a potential buyer without an operations manual. The knowledge needed on a day-to-day basis to make the business functional and successful is the true value of a company, but it is often only retained in the heads of a few people.

The Three Steps of Tactical Tasks

The tactical tasks (unpredicted outcomes) have three primary steps: 1. Ensure your team is adequately trained. 2. Make sure your team understands the Commander's Intent. 3. Empower your team to make decisions. Let's first break down the first step.

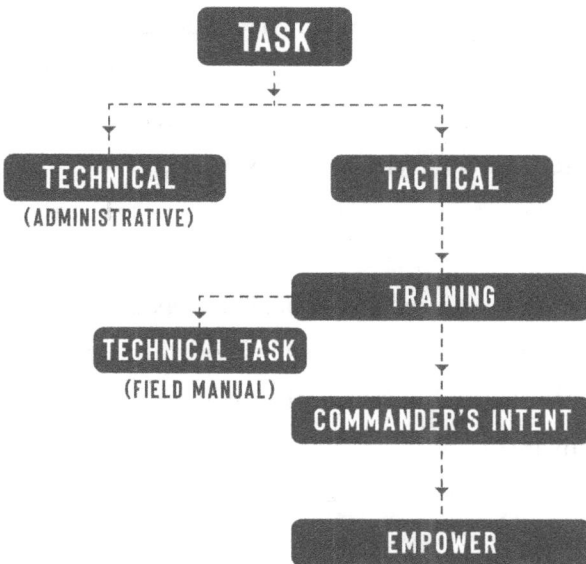

```
                        TASK

        TECHNICAL              TACTICAL
      (ADMINISTRATIVE)

                                 TRAINING

              TECHNICAL TASK
               (FIELD MANUAL)
                            COMMANDER'S INTENT

                                  EMPOWER
```

Figure 8-3

Training

The first step in solving tactical problems is training your team adequately. This seems like common sense, but it is far too often overlooked. You must ask yourself if your team has all the training they need to complete a task. Here's how to determine if your team has proper training.

Determine a baseline approach to completing a task or problem. What are the most common problems your team will deal with? For an infantry platoon, one common problem would be a head-on fight with the enemy. We have a battle drill to train for this. We rehearse it over and over until it becomes second nature. Regardless of how chaotic the situation becomes, you know what to do. Once we have drilled this scenario into muscle memory, we add more complex variations (which brings us to the next step).

Anticipate variations. For example, what if the platoon gets attacked from multiple locations simultaneously? What if the enemy does x, y, or z?

Now your team needs to **rehearse, rehearse, and rehearse.** Get as many reps as possible. This way, if they must deal with a complex tactical problem, they will hopefully have trained in a similar situation at some point. In the worst-case scenario, they will have well-practiced baseline battle drills to fall back on.

The next step is for the team and leadership to **assess the strengths and weaknesses** of the group. I will discuss this in greater detail later in the book.

Use this assessment to **design a training plan to address deficiencies**.

This entire process is ongoing. You are constantly assessing training, developing training plans, and practicing.

New people are coming into organizations all the time and it may often feel like you're backtracking, but it's part of the process. The new people need to be trained as much as possible in different scenarios they might come across. In the military, we train for every possible response the enemy may have. If we advance and they retreat, we will pursue them. If we pursue, and it turns out they were baiting us into an ambush, then we will practice responding in that scenario. If the enemy flanks, we will prepare to hold key terrain and call aircraft to control their movement. We rehearse every strategy, so we have something to reference anytime we get into a comparable situation. The training plays out in our heads, guiding us about how it was done before, how the enemy reacted, and effective ways to respond. In times of crisis, we can always turn to training.

Empowerment

The last part of dealing with tactical problems is empowering the person on the ground to make decisions. This reinforces the importance of knowing the Commander's Intent. We empower the soldiers—or the team members—by making sure they understand what needs to be done. If an individual understands the big picture and the desired outcome and is also trained to react in diverse situations, they will be successful. They will always keep in mind what the outcome needs to be.

The leader will not always be there to guide the team. There may be situations in your work where the response window is narrow, and it falls upon the employee to judge responsibly. Therefore, we must empower and equip the entire team to make decisions. This empowerment comes from training, knowledge, and confidence. It builds a kind of trust in the team that is directed from everyone, toward everyone. Trust makes the team stronger. This is true for most relationships, you trust people to make the right decisions. Why? Because you trained them that way.

As a leader and responsible team member, you lead your subordinates through diverse scenarios. As such, it is your duty to equip your team with a functional system and training. Once they're trained, they're empowered to make fruitful decisions for the organization.

Decision Making Issues

I see many leaders at every level have difficulty because they try to solve a technical problem with a tactical solution or vice versa. Being overly constrained by technical processes when you need to navigate through changing circumstances in the field can be a grave drawback. It is impossible to be effective if you are constrained by a step-by-step method. It takes away the flexibility needed to adapt to a changing enemy.

The same applies when using a tactical problem-solving method to a technical problem. Technical problems are repeatable administrative-type problems. It may seem like you're giving power to your team by telling them to figure it out, but you're wasting their time by having them reinvent the process every time. Repeating all possible solutions until you reach the right one is not the most efficient way of accomplishing tasks. On the contrary, staying organized and having detailed charts can lead to a clear understanding of the problem. Having a plan of action in which everyone knows their job eliminates the chances of wasting resources, time, and energy. Problems and issues must be identified as either tactical or technical and then dealt with by a replicable action plan.

From Reaction to Proactive

Anticipating problems before they arise is possible through experience and learning. When you can identify the best approach for each type of task, you will be able to become proactive instead of reactive, and the most challenging aspects of your job will become second nature. You have undoubtedly been in a position where you are constantly reacting to whatever gets thrown at you. However, by doing routine things routinely and establishing a training plan for educating team members, you will stay ahead of the curve.

You and your team will feel confident, well-equipped, and powerful. You will be ready for anything and, with muscle memory, able to conquer obstacles.

KEY POINTS

- Decide if the task needs routine or creativity—then handle it accordingly.

- Lock in efficient, clear methods for repetitive tasks—don't waste time.

- Make sure your organization is structured and roles are clearly defined.

- Train your team and empower them—give them the skills and freedom to get it done.

9
Reflect and Improve

"Without reflection, we go blindly on our way, creating more unintended consequences, and failing to achieve anything useful."

— MARGARET WHEATLEY

ASSESSMENT IS A CONSTANT AND ONGOING PROCESS. It is something we need to do daily as individuals and as a team. As an individual, are we truly following the leadership principles necessary to build the best team possible? We need to reflect and assess to improve and maintain the standards of the team and organization. We assess so that we can change for the better. If you aren't willing or open to implementing change, then don't bother with an assessment in the first place. Good leaders and organizations are organic, not static. Change is the key to the growth and health of a team. So, we should begin by assessing ourselves. We also need to reflect on our organization as a business, team, and as individuals in that team. All of these are interconnected.

Start With Self

Before you go to bed at night, look at the person in the mirror and be honest with yourself about whether that person is doing everything possible to become the best version of yourself. Is that person the example of what they expect from their team? Or are you demanding something you aren't even doing yourself? David Goggins, author of the book *Can't Hurt Me*, refers to this as "the accountability mirror". It's a genuinely humbling practice. What are you doing as an individual that shows your commitment to living by the leadership principles discussed in this book? My recommendation is to read through chapter 10 of this book often. Make a conscious effort to do the best you possibly can to become a better leader. Determine what you've been doing well and where you can improve. Make an extra effort to apply one of the nine principles for a day or even a week. If it feels overwhelming, just focus on one method at a time. This is particularly important if you have identified an area of weakness. Don't avoid an area just because it's difficult or you don't enjoy it. If it helps, set aside a specific time block in your day to ensure it gets done. If you need to get to know your people better, then get into the trenches and get to know them. Try your best to build strong ties and figure out the issues they face. Have meaningful conversations and be there for them. Don't find an excuse to do something else. This is your self-assessment, your reflection in the mirror. Are you doing the work that needs to be done to make your team the best?

Self-Reflection: The Way to Awareness

Awareness is crucial for leaders. It's gained through self-reflection and involves recognizing and understanding one's thoughts, emotions, and behaviors and how they align with one's core values. Being mindful of one's internal state and its impact on others is important for growth and effective decision-making. You must understand not just what decisions you have made but also why you are making them.

Not only has self-reflection been taught by the wise of the world, but it is also part of many cultures and faiths. Socrates developed his philosophy of self-reflection during a challenging time in his life. He was on trial for charges of corrupting the youth and disrespecting the state religion. Despite this, he believed that the pursuit of self-examination and introspection was vital for a fulfilling life. In his moment in front of the court, he uttered his golden dictum that "the unexamined life is not worth living".

Self-reflection includes the importance of autonomous thinking so that you can utterly understand the motivations behind your actions and make choices that align with your personal beliefs and principles. By examining your experiences, both good and bad, you can learn and grow as a person, rather than simply going through the motions like a passive participant. In essence, this philosophy encourages individuals to take control of their own lives by reflecting on their experiences, questioning their actions, and continually striving to improve their character.

Japanese culture holds self-reflection as a fundamental principle to be performed in all aspects of life. This philosophy is called *Hansei*, which means introspection or self-reflection. *Hansei* emphasizes the importance of humility in learning about ourselves, our processes, and our practices so we can continually grow and improve. Kaki Okumura, a writer and lifestyle influencer, described *Hansei* as a way of questioning our assumptions and granting ourselves the power to be better.

Hansei is widely taught in schools in Japan as a continuous process that is considered a fundamental part of personal and social development. When mistakes are made, people in Japan take responsibility, and then propose a solution to prevent the same mistake from happening in the future. *Hansei* is not just practiced after failures but also after successes. The disciplined use of *Hansei* drives continuous improvement. In business, *Hansei* is a rigorous review process that helps organizations and individuals refine their practices and achieve their goals.

As you can see, the significance of self-reflection is weighty. Its importance extends to your organization and teams. Assessments can be an ordeal in and of themselves, particularly if you have never carried one out. In the next section, I will give you an overview of assessment methods and where they are most useful.

Assessing the Team

Assessing your overall organization and team is as equally important as self-reflection. To assess your organization from a leader's perspective, you need to ask questions such as: *What do you believe are the shortcomings and strengths of the team? Are milestones being met, and goals being achieved, and if not, why? Do the issues your organization is facing stem from a lack of process or is it a motivation problem?* Then get a collective assessment of the organization by getting input from everyone about how they believe the team is doing. Asking the same questions you asked yourself.

One assessment method is to use a **formal anonymous survey**. This is typically done in the form of a written survey asking questions about the current state of the organization. The military refers to this as a "command climate survey". The frequency of a formal survey varies from team to team and you will need to determine what best suits you. I would generally recommend no more than one a year. Be sure to ask questions designed to be constructive, not to degrade, or to become a forum for complaints.

It's important to know that getting an accurate assessment from your team is nearly impossible when you are the boss. Imagine your boss's boss asking you how you feel the team is doing. Saying something bad could make you suffer repercussions from your boss. You may refrain from giving an honest opinion. It may surprise you to learn how differently your team sees the issues they are facing when they are able to speak freely.

You can use formal anonymous surveys for organizational assessment as well, but informal observation methods are even more effective. An **informal observation** is getting off your ass and spending time with the troops. This includes talking to them on an individual basis and allowing them to see and collaborate with you in person. The more they see you getting your hands dirty, the more open they will be to you. After all, you're not above any task that you ask someone else to do. While you are doing tasks together, ask questions. Assuming you want honest feedback, ask: *What would you change about the company if you could change anything? What is your favorite part of the job or least favorite part of the job? Are you getting the resources needed to do the job? Do you feel like you're informed about what's happening and why?* Do not ask them to throw their supervisor or team under the bus. It's counterproductive and you will lose their trust if it feels like you're trying to get them to undermine others. Be open to hearing anything, but keep it productive. And don't make it a time for you to make excuses or over-explain and justify why something is the way it is. This is your time to listen and learn.

The After-Action Review

In the military, we do an "After-Action Review" immediately after any training event or operation. At the very minimum, we cover what was supposed to happen, what did happen, what we need to improve on, and what went well that should be sustained. The aim is to always end on a positive note with a sustainable approach. The After-Action Review, or AAR, is a crucial process that evaluates the effectiveness of an operation, project, or process. It serves to identify what was successful, what could have been improved, and what lessons can be learned for future situations. This discussion should be clear, concise, and objective, focusing on the facts. The purpose of the AAR is to continuously improve processes and operations, and it is important to be thorough and honest in the evaluation to achieve this goal. This review always happens regardless of the size of the operation. When you can become efficient at moving three thousand troops and tens of millions of dollars of equipment to the opposite side of the world on a forty-eight-hour notice, I assure you, a constant review of every aspect has been thoroughly dissected, discussed, and rehearsed.

The most basic version of an AAR that you can apply is what we call the "three up, three down", which means naming three things to sustain and three things to improve. This could be carried out individually or as a team. It forces individuals to think critically about their performance. Be honest about your role as a leader in the success or setbacks of the team. This review is about everyone, not just subordinates. After all, you are a team. Your team will appreciate seeing you identify things you need to improve. Make sure it doesn't become a bellyaching session about what everyone else needs to do better. This is where the accountability mirror is important.

What do you need to improve on and what went well? Take the appropriate action.

Without assessments, you and your team will continue to make the same mistakes. It's important to be a team that self-corrects, always evolving and improving. You should also focus on improving your work culture, mission, and incentives, not just when you are struggling but also when the business is booming. Assessments are not supposed to be a burdensome activity, but rather something you and your team look forward to so you can improve. It should be considered a gateway to success and problem management. For instance, your company might be experiencing a communication gap that has gone unnoticed by leadership. However, when conducting an annual evaluation survey, it becomes apparent through employee feedback that this is the issue. The after-action review is a diagnosis of issues that your company might be facing. Once identified, you resolve those issues and move ahead with other problems.

The Motivation Triangle

I often find the primary question that leaders need to ask themselves is: *Do these people really want to be here? Do they want to be part of this team?* The answer to this question determines the amount of energy and motivation they are willing to contribute. Are they going to strive to excel, or will they do the bare minimum? When assessing the motivation of your team, you need to analyze a few key components. I call them **the motivation triangle**. The motivation triangle describes the three critical areas for providing fulfillment in your team. These three key areas are how people derive satisfaction from what they do. These three areas are:

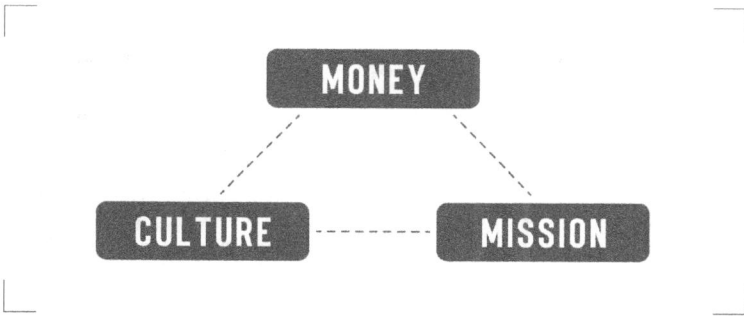

Figure 9-1

Money, Mission, and Culture.

These are not in order of importance as they will have different values for everyone. An important thing to consider when analyzing this is that there absolutely must be a high ranking in at least two of these categories to be successful. Obviously, all three would be ideal and would most likely create a bulletproof team.

Money

The more financial incentive there is to do a job, the more people are willing to do it. This incentive isn't always cold, hard cash. It also includes a 401k, health insurance, bonuses, and even a company car. Like all incentives, money is different for everyone. Some people place more importance on financial incentives than others.

In many cases, we are often limited in how much money we can offer our team members. We might not have any influence on the budget at all. As far as the military goes, we have absolutely zero say in how much people get paid. In fact, how the strictly set pay scale is a sticky point for many in the military. Two people of the same rank with the same amount of time in the military are going to get paid precisely the same amount but could be doing very different jobs of varying difficulty. Also, regardless of how good they are at their job or how much effort they put into their work, they are paid the same. It's difficult to explain and impossible to justify the reasoning. I've been on jobs as a contractor where everyone made the exact same amount regardless of experience or time on the job. Whether fifteen years of experience or five months, they were paid the same. If pay is out of your control, then you must find creative ways to make the other aspects of the triangle even more appealing. You need to enhance the culture and mission. Think about the percentage of people on your team who would quit right now if an opportunity came along to make slightly more money. How could you change the mission or culture of the team for the better? Imagine if you had an environment where everyone stayed by choice.

When the pay for your team members is "industry standard" or "competitive", then you need to upgrade it. If you're paying nearly the same as your competition, then your team can go anywhere and get the same pay they are earning from you. You need to be, as much as possible, more than competitive in your pay scale. Don't get by with only two pillars of the triangle to avoid paying your team the best you possibly can. "I'll pay a little less to save some bottom line and just have an outstanding culture." No! It's bullshit. Why do we skimp when it comes to the most important assets? It's like shopping for the cheapest food and wondering why we're not performing at our peak. Pay more and get the best quality food you can. Pay more and attract the best team you can. Don't be a cheap leader. It also lets your team know where your priority lies.

Work Culture

Culture's influence on people's actions is truly remarkable. Throughout numerous conversations with veterans, I specifically inquired about their reasons for joining the military. While some cited practical motivations like college funds, job security, or a lack of other options, others expressed a profound sense of duty to serve their country.

However, the most intriguing insight came to light when I looked into why they continue to risk their lives in combat. At this point, monetary incentives and personal gains fade away, and a resounding theme emerges: they do it for the individuals beside them, their comrades in arms. A culture within the ranks is born out of difficult circumstances.

In such moments, the mission and culture converge seamlessly. The mission transforms into a collective determination to accomplish the task at hand and ensure the safe return of everyone involved. This exemplifies the power of culture in forging unwavering camaraderie and unity among those facing adversity.

This example illustrates how culture can drive people to extraordinary lengths and inspire them to put the well-being of others before their own. Whether in the military or any setting, a strong and positive culture can foster a shared sense of purpose and motivate individuals to overcome challenges together. In essence, culture forms the bedrock on which missions are accomplished, bringing people together and binding them to a common cause.

People who respect or enjoy their work environment will be more motivated to show up and do the work. Easy concept. If they enjoy the people they work with, they will feel a sense of responsibility to their peers to get the job done and pull their weight. They will also be more likely to help others when needed.

Culture is composed of many elements, including a sense of belonging. A friendly and fun atmosphere where people can be themselves. Good working hours. Work trips, vacation time. This is also where a work-life balance fits in. I'm sure you'll find certain people will become consumed by their work just as I have. While this can be a good indicator of a healthy triangle, it is essential not to let your team burn out. This is another way to set the tone mentioned in Chapter 3. Be the example of a work-life balance. Or at least be realistic in your expectations. Just because you live, eat, and breathe your business, doesn't mean your team has to do the same.

An excellent work culture fosters collaboration, communication, respect, trust, and transparency among team members, resulting in higher levels of morale, job satisfaction, and productivity. The presence of an excellent work culture also serves as a draw for top talent, as people are more likely to be attracted to an organization with a supportive and inclusive environment.

Open communication is a key aspect of a positive work culture, allowing employees to freely share ideas and contribute to the company's success. Clear and attainable goals and a sense of community also motivate and engage the team, leading to not only benefit the team but also have a positive impact on the organization. By prioritizing positive values and behaviors in the workplace, companies can create a dynamic and successful environment that supports growth and success in the long term. This sort of dynamic culture can be created by following the principles and tips in this book.

Be a visionary. It's about being optimistic and the type of leader who looks for opportunities and is willing to incorporate new ideas. Embrace possibilities and be agile enough to shift gears when the situation demands. Getting stuck in old habits? Adaptability is your best ally—be it embracing new tech or fresh ideas.

We all love our comfort zones. They're cozy and familiar, but they're also a place for stagnation. This isn't just about pride, it's about survival on the battlefield. When the enemy's or competitor's tactics evolve, you can't be the last one to catch on.

Your team is watching you, always. They're not just looking for a leader, they're looking for a beacon of progress. If you're not pushing boundaries, evolving yourself and the organization, don't be surprised if your team's eyes start wandering, always on the lookout for the next big thing. They'll follow a leader who's always one step ahead to the ends of the earth.

Be a shield. You've absolutely got to have their backs! When you're leading a team, they need more than just orders, they need a shield. They want protection when under your command. Your troops should feel that unshakeable confidence that you're not just their commander, but their fiercest advocate. In addition to giving commands, be the first to jump into the fray for them when the chips are down. When you have their backs, they will want to have yours.

No one wants to be left out in the cold when things go sideways. This includes when they make a mistake. Your team's loyalty hinges on this unwavering belief that you won't leave them high and dry when the going gets tough. Not just in the heat of the moment. They need to feel rock-solid about the future too. If your team's always thinking their jobs are only as stable as a house of cards, you've already lost them. They'll have one foot out the door and looking for the next haven.

Mission

If your team feels part of something bigger than themselves, they will more easily connect with the mission. They will see the results in the world and how it produces something positive. The mission doesn't have to be solving world hunger to get a sense of accomplishment and feel the drive to achieve it. A sense of purpose can come from many things.

Chances are your product or service helps save someone time or even solves their problem. Remind your team of the mission and their crucial role in achieving it. Each day, they should understand that their efforts make a difference for others.

If someone is a subject matter expert in their field, they will feel a sense of purpose. Make each person feel like a go-to resource for information. It's rewarding to be needed and recognized as an expert. Provide the training needed so they have something significant to contribute to conversations. Turn them into subject matter experts.

The more knowledge they have, the more inclined they'll be to contribute. Educate your team on the background and history of the field, as well as the business aspects and how these impact the bigger picture. Discover what each individual wants to learn about their specific role or something broader within the company. Make them feel like you are investing in their future.

If people are paid enough and believe in their mission, they can generally tolerate a less-than-ideal culture. They will be able to mentally disconnect and get that fulfillment someplace other than work. On the other hand, if a workplace has a great culture and a fulfilling mission, most people are willing to deal with slightly less pay. You get the idea. The absolute best is to have the trifecta. Pay them well, create a fun and inclusive culture, and have a powerful sense of mission. With all three, you will exceed any expectations.

Remember, as leaders, you need to reflect on and review yourself, the team, and the overall mission, project, and productivity. You also need to assess and reflect on your team's climate—how happy, motivated, and fulfilled they are. Your team will only trust you and work honestly if you make them feel comfortable, relaxed, and trusted. You must honor the work of each team member and ensure they know it.

If you have not assessed your organization yet, then it is time for you to design an assessment method that best suits your work conditions. Trust me, it will open new doors for you and your team. And if you are struggling with the incentives and what to look for when assessing, then remember the trifecta: money, mission, and culture. These are three magical words of organizational assessment. The goal is to always reflect and improve. I can assure you these aren't just ideas to use if the mood strikes you. These are all rules you should be implementing religiously. If you aren't doing assessments, you are missing a key component to making your team as effective as possible.

KEY POINTS

- Look in the mirror—self-awareness is the foundation of strong leadership.

- Give your team solid feedback—stay open to criticism yourself.

- Run After-Action Reviews often—always reflect and improve.

- Get a grip on what motivates your team—money, mission, or culture.

10
Get Started

"You can't build a reputation on what you're going to do."
— HENRY FORD

One of the most significant differences between a good leader and a great one is the ability to inspire others. To become an inspiring leader you must exude an energy that commands attention as soon as you walk into a room. This requires a blend of charisma, energy, positivity, and offering authentic care and respect to those around you. Embrace accountability for yourself and others, while simultaneously nurturing optimism. Being a great leader is about spotlighting the positive and relentlessly seeking solutions instead of getting bogged down by problems. Remember, your optimism isn't just for you; it can ignite inspiration and spark positive change in those around you. The leaders I remember the most are the ones that inspired me.

Absorb the persona of leaders who've paved the way before you and make leadership your craft. Seize the valuable lessons they offer and integrate them into your style. Recognize their missteps and choose a different path. Let humility be your guide. Weaving these principles into the fabric of your leadership and setting a powerful example for your team, you can foster an environment where everyone is motivated to strive for excellence.

Now that you have read the book, you have a better understanding of these pivotal leadership principles. Leadership is a practice in the art of influence, rooted in the bedrock of personal relationships. Think of this as your daily dose of power insights. Keep it at your fingertips, review it, and live it. Going from good to great happens when these principles shift from being mere ideas to actions that you personify.

Simply reading this book (or any other leadership book for that matter) is not enough. You must implement these ideas in order to affect change and growth. So, let's distill the principles down. Below are the cliff notes—the condensed wisdom you need to lead and succeed. Take these notes, choose your focus, and make each week your launchpad to leadership excellence. If it feels like too much, just zero in on one or two key insights each week. This isn't a sprint, it's a marathon. Methodically transform a thought into a habit, a habit into a lifestyle, and a lifestyle into your legacy.

Begin by having real conversations with your team. Connect with them on a personal level, beyond the workday tasks. See who each person is behind their role. When you commit to action, honor that promise. Your follow-up and follow-through are the currency of trust in your leadership.

Leadership is an ongoing journey of learning, one that requires dedication and decades to refine. I wish you luck in your journey.

KEY POINTS & TAKEAWAYS

1. Trust and Respect

- Put your rank aside and lead with respect. Treat everyone the same.

- Own your words—don't lie, don't sugarcoat, and always follow through.

- Get in the trenches and work hard—your actions show what real competence and character look like.

- Bring your team into the fight—let them own the plan and use their strengths to drive success.

- Build real relationships. Treat your team like family—lead with honesty, toughness, and connection.

- The respect and effort your team gives you is a mirror of your leadership—stay composed and balanced, even under pressure.

2. Know Your People

- Know your team beyond their names—see them as more than just cogs in the machine.

- Build real connections—care about them as much as you expect them to care about the job.

- Be approachable—make sure everyone feels comfortable bringing their concerns to you.

- Follow up—make sure your team knows they're valued and heard.

- Get out of your office—connect with your people on a real level, learn their stories, and find common ground.

- Treat every day as a chance to focus on what matters—your team is your priority, treat them like it.

3. Check your Ego

- Check yourself. Know when your ego is calling the shots and get real about what's healthy and what's not.

- Pause before you react—don't let emotions or

competition control you.

- Listen with an open mind and try to see things from the other side.

- Be confident without needing applause—don't hoard info or seek validation.

- Stay sharp—know how ego shows up in yourself and others, and don't take things personally.

- Learn to shut up when your ego wants to speak—use your head, not your pride.

- Trust yourself to know what needs to get done—or figure it out when you don't.

4. Commander's Intent

- Keep the mission clear and to the point—brevity hits harder and resonates.

- Lay out what matters most—then trust your team to get it done.

- Empower your team to make decisions—use decentralized command to keep things moving.

- Embrace failure whenever possible—it's how you build future leaders.

5. Delegate to the Lowest Level

- Get some weight off your plate—delegation isn't dumping your crap, it's building leaders.

- Build leaders every day—don't wait for a crisis to develop them.

- Know your team's strengths and needs—mentor them with that in mind.

- Push your team to delegate as well—empower others to step up.

- Pass authority but stay responsible—make sure they've got what they need to succeed.

- Use drills like "one-up" to get people thinking beyond their roles and prepare them for leadership.

6. Watch Your Tone

- Lead by example—your behavior sets the tone for the whole team.

- Speak positively about your job and your team—don't allow a culture of negativity to take root.

- Respect everyone like they're family—cut the demeaning crap.

- Keep it real—speak honestly and upfront; talk to others like you want to be talked to.

- Remember, tone is amplified under pressure—practice responding effectively in high-pressure situations to maintain control.

7. Rewards and Punishments

- Reward good behavior every day—show genuine appreciation for the effort.

- Put real thought into recognition—make it meaningful and tailored to what drives them.

- Don't reward bad behavior or hand out awards like candy—make recognition count.

- Be fair and impartial when it's time to punish—don't play favorites.

- Be thoughtful with corrective actions—consider their personal struggles when handling mistakes.

8. Find Structure

- Decide if the task needs routine or creativity—then handle it accordingly.

- Lock in efficient, clear methods for repetitive tasks—don't waste time.

- Make sure your organization is structured, and roles are clearly defined.

- Train your team and empower them—give them the skills and freedom to get it done.

9. Reflect and Improve

- Look in the mirror—self-awareness is the foundation of strong leadership.

- Give your team solid feedback—stay open to criticism yourself.

- Run After-Action Reviews often—always reflect and improve.

- Get a grip on what motivates your team—money, mission, or culture.

Acknowledgements

There have been so many people who have inspired, mentored and motivated me along my journey.

To my wife, Andrea: You are the most fantastic person I know. Thank you for listening to my tireless talks about this book and for being an ear for me to clear up my ideas. You were my best editor, designer, and sounding board. You were also my most insightful critic, even though I often pushed back. Your unwavering support and belief in me kept me going, even when the writing process felt overwhelming.

To my kids: Thank you for understanding my need to write this book and for being patient with the time it took to accomplish my goal. I hope this book serves as a testament to the importance of following one's passion.

To my close friends:

Patrick Heringer: Your encouragement and camaraderie have been invaluable. You've always been there to offer a laugh or a word of wisdom when I needed it most. Our conversations have greatly influenced my understanding of where I want to be in life.

Tony Salinas: Thank you for your constant motivation and for pushing me to be a better writer. Your mentorship and friendship have been a rock for me throughout this journey.

Florent Groberg: Your perseverance and strength have inspired me more than words can express. You are the epitome of what we strive to be. Thank you for all you do to represent our brotherhood and for ensuring the world knows the caliber of the men we served with.

To some great men who sacrificed everything and are gone too soon: Ron Ogle, John Wade, Ryan Rojas, Mike McNulty, Darrell Griffin Jr., Patrice White. You are gone but never forgotten. Your memories continue to inspire me every day. This book is, in part, a tribute to each of you and the lasting impact you've had on my life.

To my editor, Carmen: Your attention to detail and insightful feedback helped shape my ideas into something cohesive and impactful. I am grateful for your expertise and hard work.

Lead The Way!

About the author

Korey Staley was born and raised in Atlantic, Iowa, where he enjoyed all things outdoors and was active in many sports. After finishing high school, he did what many young people do, he joined the Army. For the next twenty-two years, Korey served his country as an Infantryman. He was decorated with several Bronze Star Medals with Valor and earned qualifications such as Airborne, Ranger, Sniper, and Drill Sergeant, just to name a few. Korey retired a U.S. Army First Sergeant, having directly led over 600 soldiers, including over three years leading in combat. This exceptional career showed him unique insights into leadership during extreme circumstances, prompting him to write his first book, *The Fieldcraft of Leadership*.

Korey believes in passing on knowledge so future generations can benefit from the experience and hardship of those who have gone before.

He currently resides with his family in Spokane, Washington. He is a contractor to the U.S. Department of Defense for the SERE (Survive, Evade, Resist, Escape) course and for The Center for Personal Protection and Safety (CPPS) as an Active Assailant and High-Risk Travel instructor.

Korey has been cited, discussed, and consulted in numerous books to include:

8 Seconds of Courage: A Soldier's Story from Immigrant to Medal of Honor by Florent Groberg

Siren's Song: The Allure of War by Antonio Salinas

Last Journey: A Father and Son in Wartime by Darrell Griffin Sr.

To War with the Fourth: A Century of Frontline Combat with the U.S. 4ᵗʰ Infantry Division, from the Argonne to the Ardennes to Afghanistan by Martin King, Jason Nulton, and Mike Collins